P9-DCH-826

FRAGILE EARTH

Collins

Fragile Earth

Collins
An imprint of HarperCollinsPublishers
Westerhill Road
Bishopbriggs
Glasgow
G64 2QT

First Published 2006
Second Edition 2012

Printed in China by South China Printing Co. Ltd.

British Library Cataloguing in Publication Data.
A catalogue record for this book is available from the British Library.

ISBN 978-0-00-745523-2
Imp 001

Front cover photograph: Puyehue – Cordón Caulle volcano, Chile © Reuters/Carlos Gutierrez
Front flap photograph: Tsunami swirls in Oarai, Japan, AP/Press Association Images
Back cover photograph: Gulf of Mexico oil spill, MODIS/NASA
Back flap photograph: Rikuzen-takata, Japan, JIJI Press/AFP/Getty Images

All mapping in this book is generated from Collins Bartholomew digital databases. Collins Bartholomew, the UK's leading independent geographical information supplier, can provide a digital, custom, and premium mapping service to a variety of markets.
For further information:
Tel: +44 (0) 020 307 4515
e-mail: collinsbartholomew@harpercollins.co.uk
or visit our website at: www.collinsbartholomew.com

If you would like to comment on any aspect of this book, please write to
Collins Maps, HarperCollins Publishers, Westerhill Road, Bishopbriggs, Glasgow G64 2QT
e-mail: collinsmaps@harpercollins.co.uk
or visit our website at: www.collinsmaps.com
Follow us on Twitter @CollinsMaps

CONTENTS

INTRODUCTION

One thing we can be certain of is that change will happen. The nature and extent of environmental change around the world is explored in great detail through this book and it raises many issues relating to the effects such change can have and how it can be managed, minimized and controlled. The impact of change is clear for all to see through dramatic images illustrating the extent of natural and man-made disasters, the effects of development projects and the results of long-term natural processes.

Images, statistics and maps throughout the book point to how these changes affect people and to the serious issues faced by some of the world's most vulnerable people as a result of events beyond their control. The inter-relatedness of the phenomena portrayed is also clear – earthquakes can trigger further catastrophe; global warming and the resultant shrinking of glaciers and ice sheets can contribute to more frequent earthquakes; the growth of cities can put great strain on water resources and ecosystems. Although man is behind many of the changes, and appears in some cases to be making their effects worse, it seems to be man's ingenuity, creativeness and collective will that are coming to the fore in finding solutions and bringing hope.

Recent events highlighting the ferocity of the Earth's forces and the natural disasters which can result from them are dramatically shown in the first chapter. It includes images of recent earthquakes in Haiti, New Zealand and Japan which caused widespread devastation. The Japanese earthquake triggered a deadly tsunami causing extensive damage to coastal settlements and to the Fukushima nuclear power plant, while volcanic ash from Icelandic volcanoes caused travel disruption across Europe. Such events remain impossible to predict and it is difficult to forecast just what effects they may have.

Chapter 2 focuses on storms, tornadoes, and other extreme weather events which, while perhaps easier to forecast and monitor than earthquakes and volcanoes, still cannot be directly controlled. It is not clear whether man is influencing changing weather patterns, but it seems that the frequency of particularly destructive storms is increasing, posing a particular threat to heavily populated coastal areas in those parts of the world where tropical storms regularly occur – compelling images of the widespread effects of a cyclone in Myanmar highlight this.

More obviously within man's control are issues presented in the following chapter: the growth of cities – more than half of the world's population are now urban-dwellers – shortage of land and pressure on natural resources. Demand for land itself is evident in dramatic images of reclamation projects in the Middle East and Hong Kong but perhaps most crucial are the questions of how best to control and manage increasingly scarce water resources for a growing and demanding population. There are serious issues of both the quantity of water being consumed by people, agriculture, industry and energy generation, and of water quality – huge numbers still do not have access to safe water or sanitation. These are hugely complex issues, but a balance must be, and is being sought and while we may all contribute, albeit inadvertently, to the problems, we can all be part of the solution.

Man's impact on the Earth is starkly illustrated through the images of deforestation and pollution in Chapter 4. The inter-relatedness of issues is again evident here – for example, burning of forests for agriculture generates huge amounts of carbon dioxide, contributing to global warming. These are global issues too – their effects are very rarely just local but can have an impact far away from their origin. Hence joint efforts are taking place to address such issues as deforestation – the fascinating and shocking sequence of images of deforestation in Brazil point to the urgency of this issue in particular – and it is to be hoped that lessons have been learned from the recent oil spill in the Gulf of Mexico and the incident at Fukushima.

Global warming is still high on the agenda of environmentalists, governments and other agencies. Debates on its causes, its likely effects and the timescales involved continue, and are likely to do so for some time. While they do, and as images in the next chapter show, glaciers continue to melt, and ice caps continue to shrink – perhaps illustrated most dramatically through images of changes in the polar regions. No easy solutions are presenting themselves, but changes in behaviours, a strong global collective will and political imagination will all be required.

Vulnerability to environmental change is particularly evident in areas subject to the spread of deserts – the process of desertification. This is illustrated under the chapter 'Parched Earth' by images from the Sahara desert and the adjacent Sahel region of Africa. This process, plus the effects of drought and fire, and the drying up of rivers and lakes can have devastating effects on local communities. A huge diversity of cultures and nature are affected by these phenomena and need protection from their worst effects. Again, it is the ingenuity of individuals, societies, governments and other organizations that is looking to address these problems. Change is possible through right choices and looking beyond purely economic factors.

While our use of water presents challenges, it's inherent power can also have dramatic effects. Images in the final chapter provide clear evidence of the natural destructive and constructive forces of water along coastlines. These can cause long-lasting and far-reaching damage, particularly if nature is not allowed to keep them in balance. While the sea cannot be held back and controlled completely, imaginative solutions are being developed to minimize its destructive impact. Similarly, rivers in spate and flood can be hugely destructive and the effects of floods can change landscapes and lifestyles for ever. Pictures from recent floods in Angola and Queensland, Australia also emphasize that such events are not restricted to specific parts of the world but can happen almost anywhere.

The collection of images presented here conjure up mixed emotions – a sense of hopelessness and awe in the face of natural forces which cannot be tamed and which are certain to continue to cause death and destruction, but also a feeling that man's imagination and creativeness, if put to good use, can help minimize their effects and address the needs of those most directly affected. While many of the images have desperate personal stories behind them which shouldn't be forgotten, perhaps this book can act as a spur to helping in the search for solutions and for ways to care for our fellow man when disaster strikes.

IMAGE LOCATIONS

Locations of images found in the book, and pages on which they first appear.

Restless Earth 10–65

Extreme Storms 66–113

Man-made World 114–149

Damaged World 150–181

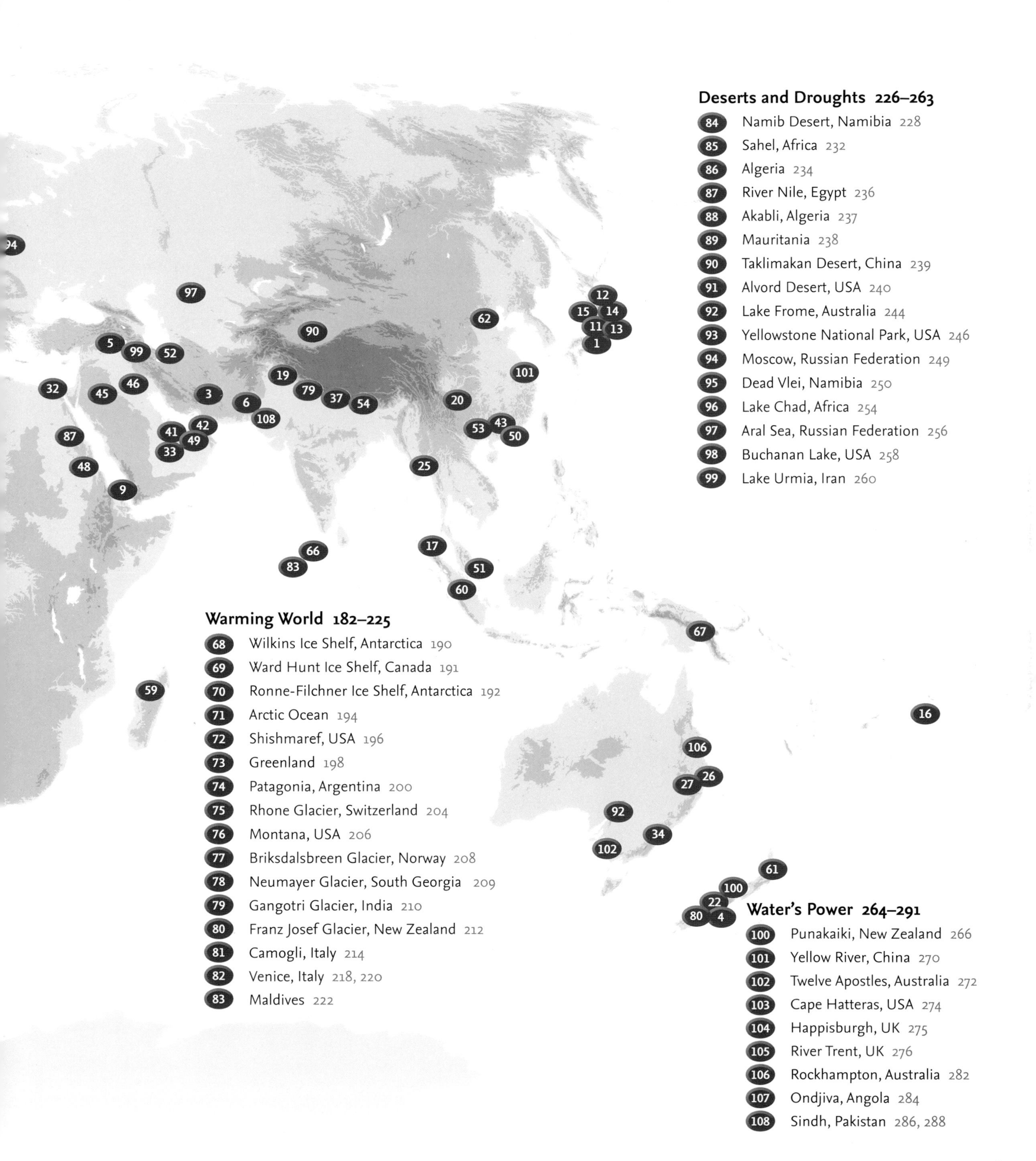

Deserts and Droughts 226–263

Warming World 182–225

Water's Power 264–291

RESTLESS

EARTH

NATURAL PHENOMENA

Earthquakes – destructive movements of the Earth's surface due to sudden releases of energy in the Earth's crust or upper mantle

Volcanoes – openings in the Earth's crust from which molten lava, rock fragments, ash, dust, and gases are ejected

Tsunamis – large, often destructive, sea waves produced by submarine earthquakes, subsidence, or volcanic eruptions

Landslides and avalanches – movements of large masses of rock, snow or ice down mountains or cliffs

EARTHQUAKES –

Destructive movements of the Earth's surface due to sudden releases of energy in the Earth's crust or upper mantle

Hanshin Expressway, Kōbe, Japan, 1995

EARTHQUAKES

The Richter scale

The scale measures the energy released by an earthquake. The scale is logarithmic – a quake measuring 4 is 30 times more powerful than one measuring 3, and a quake measuring 6 is 27 000 times more powerful than one measuring 3.

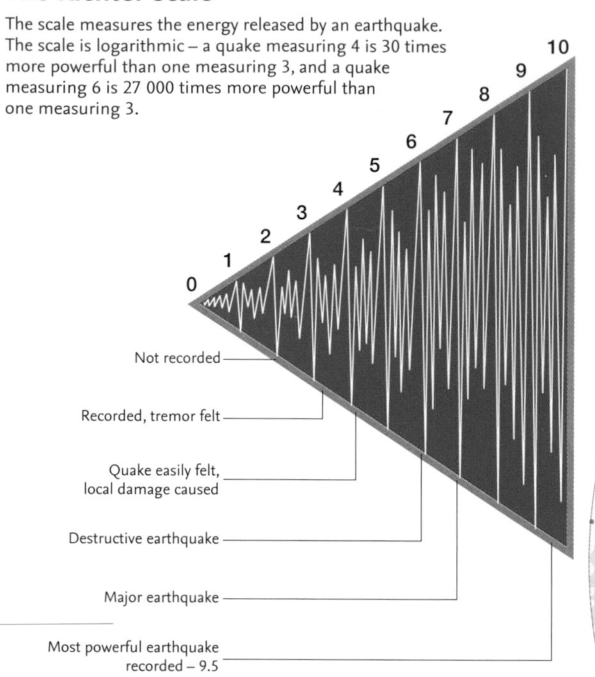

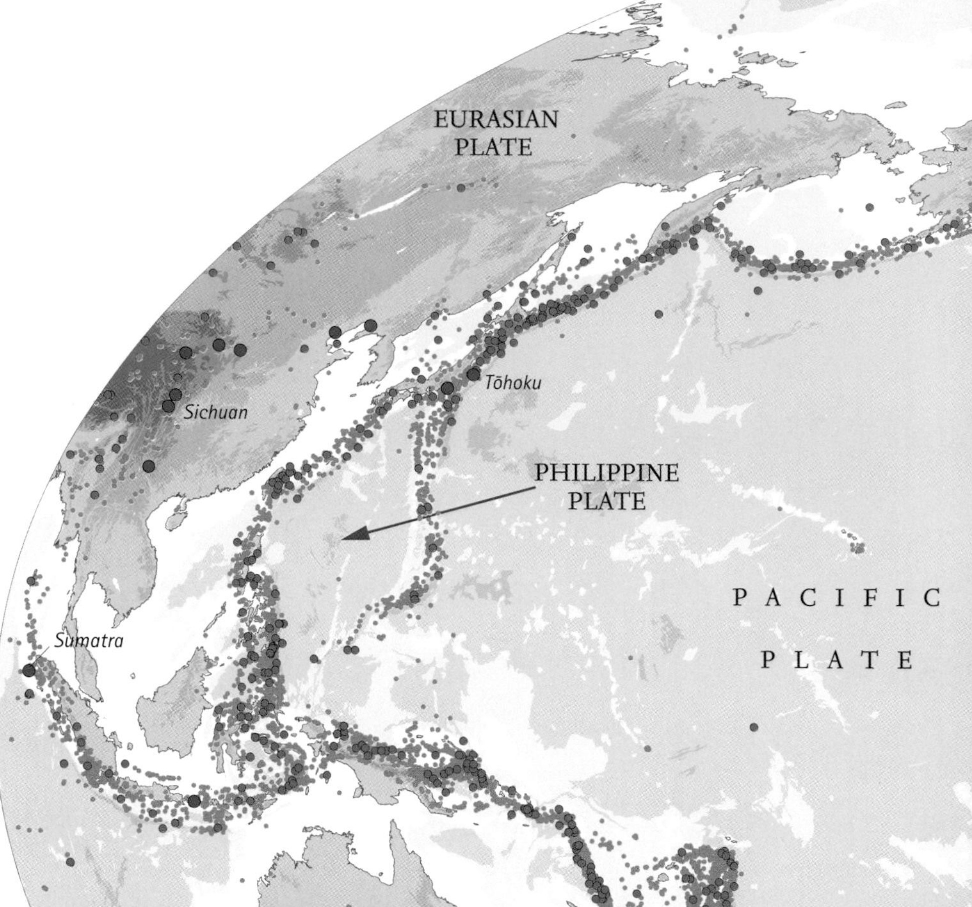

The moment magnitude scale

An alternative scale used to measure the size of earthquakes is the moment magnitude scale (MMS). It is based on the physical properties of an earthquake and specifically the seismic moment. It was developed in the 1970s to take over from the 1930s Richter scale. Like the Richter scale it is logarithmic, but has the advantage of having no upper limit to possible measurable magnitudes.

Deadliest earthquakes 1978–2011

Year	Location	Deaths	Magnitude
1978	**Khorāsan Province,** Iran	20 000	7.8
1980	**Ech Chélif,** Algeria	11 000	7.7
1988	**Spitak,** Armenia	25 000	6.8
1990	**Manjil,** Iran	50 000	7.7
1999	**Kocaeli (İzmit),** Turkey	17 000	7.6
2001	**Gujarat,** India	20 000	7.7
2003	**Bam,** Iran	26 271	6.6
2004	off **Sumatra,** Indian Ocean	>225 000	9.0
2005	**northwest Pakistan**	74 648	7.6
2008	**Sichuan Province,** China	>60 000	8.0
2010	**Léogâne,** Haiti	222 570	7.0
2011	**Tōhoku,** Japan	15 800	9.0

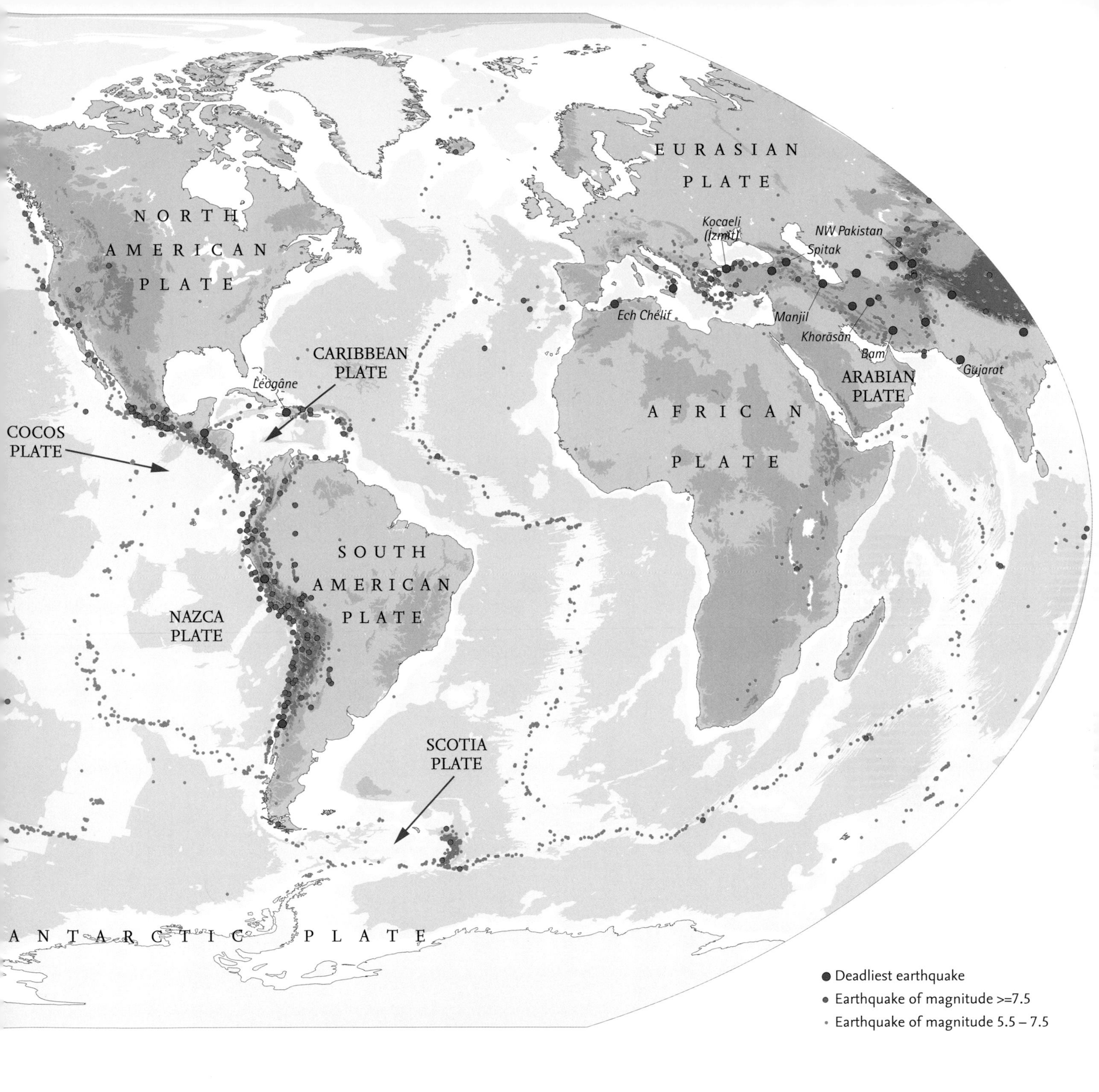

The Earth's outermost solid layer consists of a thin crust of rock. Dense rock, approximately 5 km (3 miles) thick, lies beneath the ocean basins. Lighter rock forms the continents, projecting above the ocean surface. Where there are mountain chains the continental crust is up to 60 km (37 miles) thick. Although the crust covers the entire planet, it is not in the form of a single, unbroken skin. Crustal rocks are in sections of varying size, called plates. Individual plates can move, as though jostling with their neighbours, and the boundaries where plates meet are the locations of seismic activity – earthquakes and volcanic eruptions.

Earthquakes send shockwaves through the Earth. The amplitude of these seismic waves indicates the force of an earthquake and is measured on a scale devised by Charles Francis Richter (1900–85) known as the Richter scale. The Richter scale is logarithmic, meaning that each whole number represents ten times more energy than the preceding number.

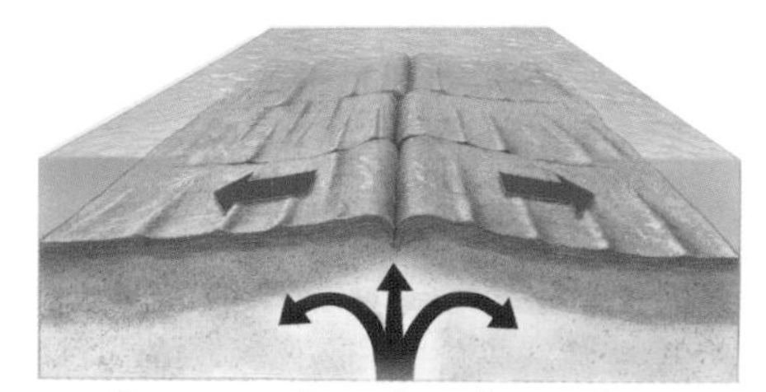

Constructive plate boundary

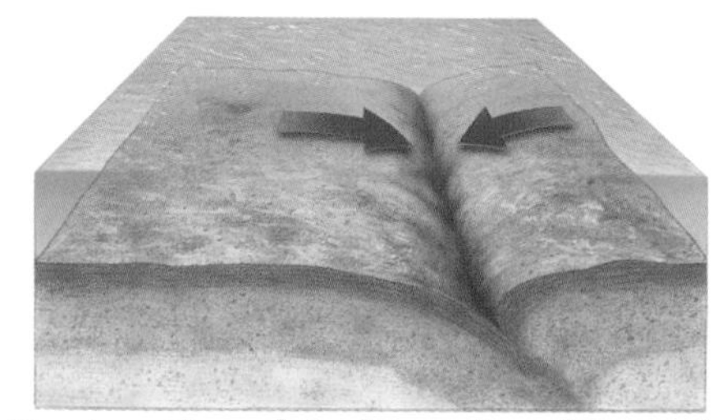

Destructive plate boundary (Oceanic)

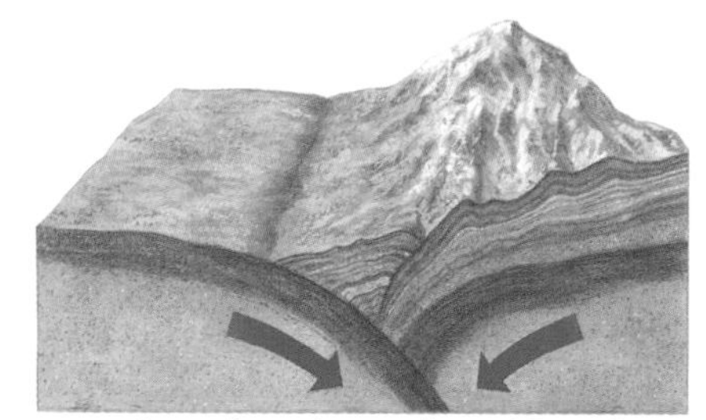

Destructive plate boundary (Continental)

Conservative plate boundary

Earthquakes of magnitude 4 or more on 11 April 2012

The earthquakes listed in the following table are located on the map above.

Order	Time (GMT)	Location	Magnitude	Depth	Number
1	00:34	Easter Island region	4.5	10 km (6.2 miles)	1
2	01:21	North Xinjiang, China	4.5	31.7 km (19.7 miles)	1
3	04:53–05:44	South Mid-Atlantic Ridge	5.1–5.2	9.7–10.5 km (6.0–6.5 miles)	2
4	06:54	Jujuy, Argentina	4.4	226.3 km (140.6 miles)	1
5	07:41	Banda Sea, Indonesia	5.3	132.2 km (82.1 miles)	1
6	08:38–23:18	off northwest coast of Sumatra	4.5–8.6	5.3–22.9 km (3.3–14.2 miles)	29
7	09:00 & 22.02	Andreanof Islands, Alaska	5.5 & 4.2	56.5 & 35.6 km (35.1 & 22.1 miles)	2
8	09:27–23:56	North Indian Ocean	4.6–6.0	9.8–15.6 km (6.1–9.7 miles)	14
9	11:28	South Sandwich Islands region	5.4	259.0 km (160.9 miles)	1
10	13:12	Tonga	4.8	245.2 km (152.4 miles)	1
11	22:41	off coast of Oregon, USA	5.9	10.2 km (6.3 miles)	1
12	22:55	Michoacán, Mexico	6.5	20.0 km (12.4 miles)	1
13	23:28	Fiji region	4.6	535.7 km (332.9 miles)	1

The Earth's plates

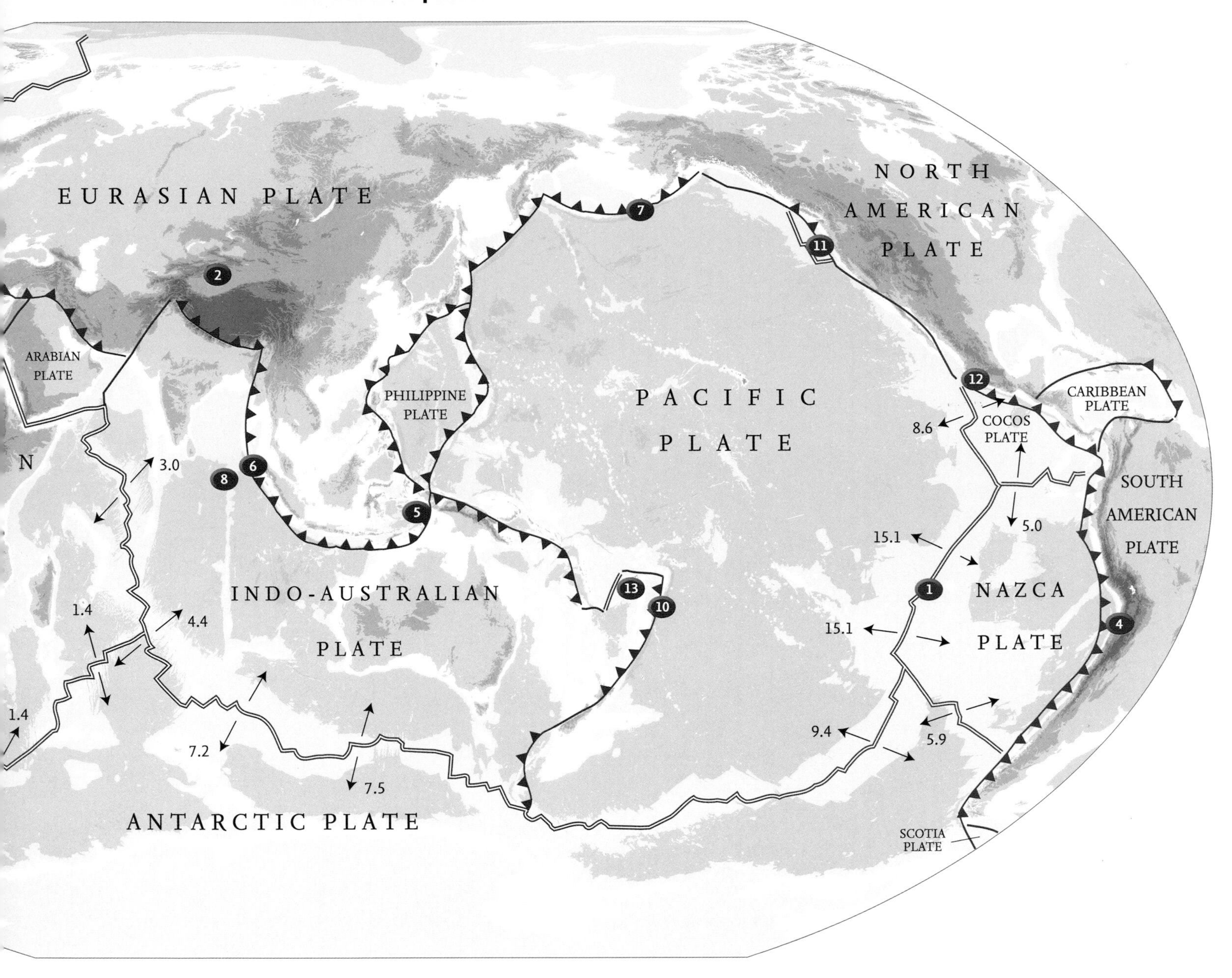

Earthquakes are common events and occur daily. On 11 April 2012, fifty-six earthquakes were recorded with a magnitude of over 4.0 on the Richter scale; their distribution can be seen on the map above. On 26 December 2004 an earthquake of a magnitude of 9.5 occurred off the northwest coast of Sumatra, resulting in over 220 000 deaths from the tsunami it initiated. Eight years later the same area is still experiencing major tremors accounting for twenty-nine of the total recorded on that day. The two strongest were recorded at 8.2 and 8.6. A further fifteen earthquakes were recorded the next day in that same area.

Plates may move away from each other, toward each other, or they may pass each other in opposite directions. Plates are moving apart at ridges along the centres of each of the oceans. These boundaries are said to be constructive, because where two plates move apart hot rock rises from beneath the crust to fill the gap, cooling to form new crust. Where two plates collide, the boundary is destructive. If both plates are made from oceanic crust, the rocks from one plate slide beneath those from the other, but drag those rocks with them. The sinking rocks are said to be subducted into the Earth's mantle. When two continental plates collide both plates sink at the base, but the surface rocks are crumpled upward to form mountains. At a conservative margin, crust is neither constructed nor destroyed as the plates move past one another, jerkily, along a transform fault.

HAITI EARTHQUAKE, LÉOGÂNE 13 JANUARY 2010

On 12 January 2010 an earthquake of magnitude 7.0 struck Haiti near the town of Léogâne. The effects of this can be seen above in the town where over 80 per cent of all buildings were affected and over 200 000 people were killed. Camps were set up in open spaces such as sports fields and food was distributed with the help of local scouts and guides. Medical facilities were set up close to the local nursing school.

HAITI EARTHQUAKE, PORT-AU-PRINCE 13 JANUARY 2010

In Port-au-Prince most of the historic centre of the city was destroyed. In the centre of the main picture is the Presidential Palace where the upper floor came down on the lower one. The effects can be seen more clearly in the before and after inset images. There is also much evidence of the destruction of other buildings in the bottom left of the image. Groups of people can be seen in many of the open spaces where they can avoid collapsed buildings, collect food and get medical attention.

BAM CITADEL, IRAN APRIL 1997

The citadel at Bam in Iran is 1000 km (630 miles) southeast of Tehran. It dates back 2000 years and is constructed mainly of mud bricks, clay, straw, and the trunks of palm trees. The city was originally founded during the Sassanian period AD 224–637 and its restoration had been ongoing since 1953. It is a UNESCO World Heritage Site due to its importance when it lay at the crossroads of major trade routes. It survived in the desert due to an efficient water management system.

AFTER THE EARTHQUAKE...

On 26 December 2003 an earthquake of magnitude 6.6 struck southeastern Iran killing over 26 000 people and destroying much of the city of Bam. About 60 per cent of the buildings were destroyed. The old quarter of the city and the citadel were severely damaged. The nature of its construction had little resistance to the force of the earthquake. There are plans to try and rebuild the citadel and some of the other buildings.

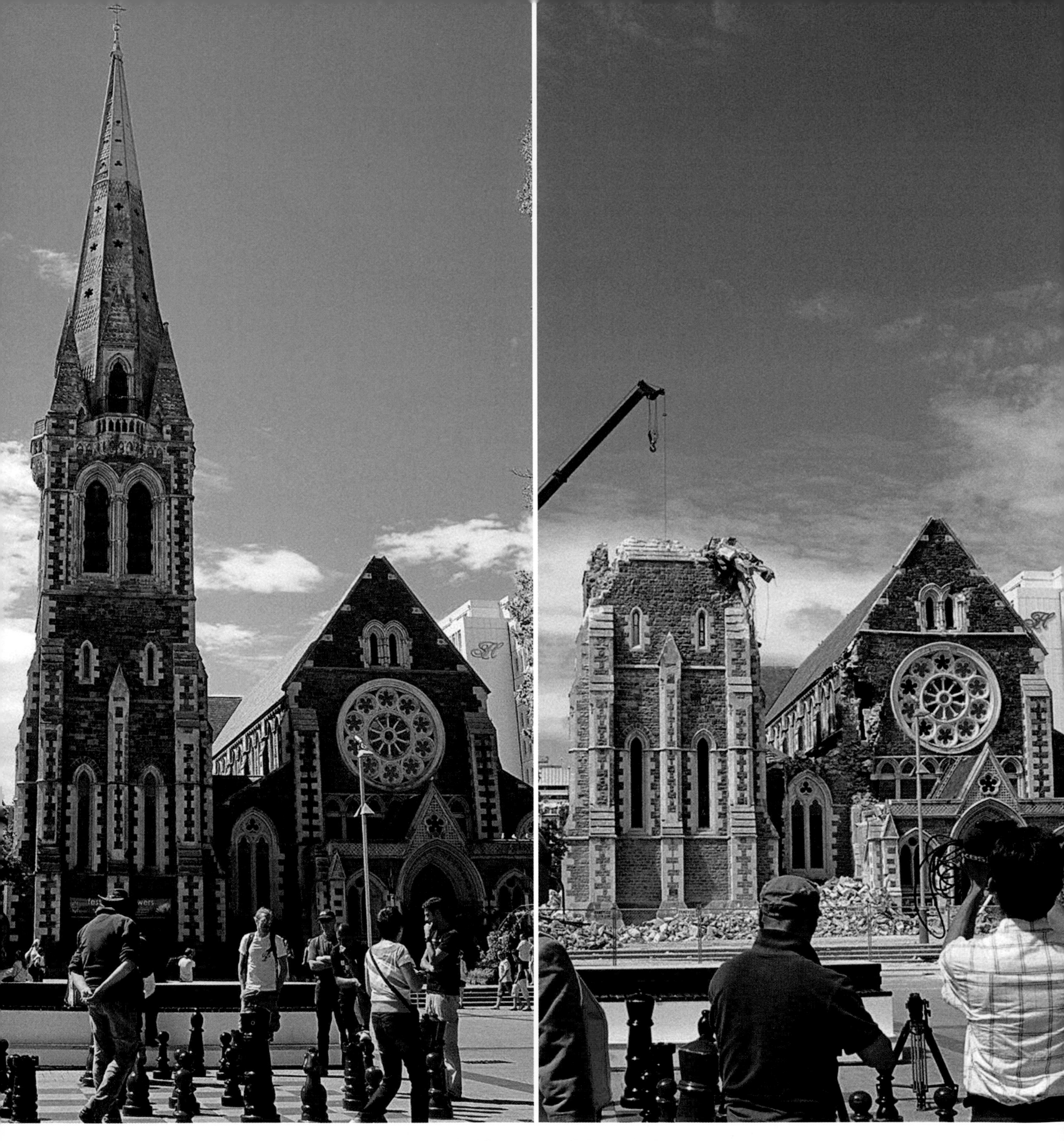

CHRISTCHURCH CATHEDRAL, SOUTH ISLAND, NEW ZEALAND 12 FEBRUARY 2011 AND 24 FEBRUARY 2011

On 22 February 2011, an earthquake of magnitude 6.3 struck the city of Christchurch, which is New Zealand's second biggest city with a population of nearly 400 000. It was the country's second most deadly natural disaster, with 185 fatalities and almost 2000 people receiving injuries. The images above show the city's iconic nineteenth-century cathedral before the earthquake and two days after. The spire was completely destroyed and the rest of the building became structurally unsound. The remains will be demolished and a new cathedral built.

CHRISTCHURCH, SOUTH ISLAND, NEW ZEALAND 22 FEBRUARY 2011

This image shows rescue workers digging through rubble trying to find survivors. As the earthquake's epicentre was just 10 km (6 miles) southeast of Christchurch and was very shallow at a depth of 5 km (3 miles), a vast amount of damage occurred in the city centre. The buildings and infrastructure were already in a weakened state from an earthquake of magnitude 7.1 which occurred six months previously. Unfortunately, the earthquake occurred at lunchtime on a weekday when the city centre was particularly busy.

VAN, TURKEY 23 OCTOBER 2011

Turkey lies in one of the world's most seismically active regions; it is located at the intersection of several tectonic plates and is crossed by numerous active fault lines. The result is frequent strong earthquakes. On 23 October 2011, the province of Van in eastern Turkey was hit by an earthquake of magnitude 7.1. The epicentre was located between the city of Van and the town of Erciş, both of which were heavily damaged. There were over 600 fatalities and more than 4000 injuries; this was largely due to the high number of collapsed buildings in urban areas.

WĀM, BALOCHISTAN PROVINCE, PAKISTAN 29 OCTOBER 2008

On 29 October 2008, numerous rural villages in the Balochistan Province of southwestern Pakistan were completely raised to the ground by an earthquake of magnitude 6.4 which occurred during the early hours of the morning. The death toll was 166, over 300 people were injured and more than 17 500 people were left homeless. A series of powerful aftershocks continued to hit the area including one of magnitude 6.4 which was as powerful as the original earthquake. The image above shows the utter demolition of the small rural village of Wām.

VOLCANOES –

openings in the Earth's crust from which molten lava, rock fragments, ash, dust, and gases are ejected

Eyjafjallajökull, Iceland

VOLCANOES

Famous historic eruptions

Date	Location	Extent
79 AD	Vesuvius, Italy	Buried the towns of Pompeii and Herculaneum and the surrounding area, firstly with ash then with pyroclastic lava flows.
1815	Tambora, Indonesia	Sent so much material into the atmosphere it created the 'year with no summer' in Europe (1816) and reduced global temperatures by 3C° (5.4F°).
1883	Krakatoa, Indonesia	The eruption only left a third of the original volcano behind, lowered global temperatures by 1C° (1.8F°), and caused unusual sunsets for three years afterwards.
1963–7	Surtsey, Iceland	A submarine eruption which created a new island of 2.5 sq km (just under 1 sq mile) which now supports plant and animal life.

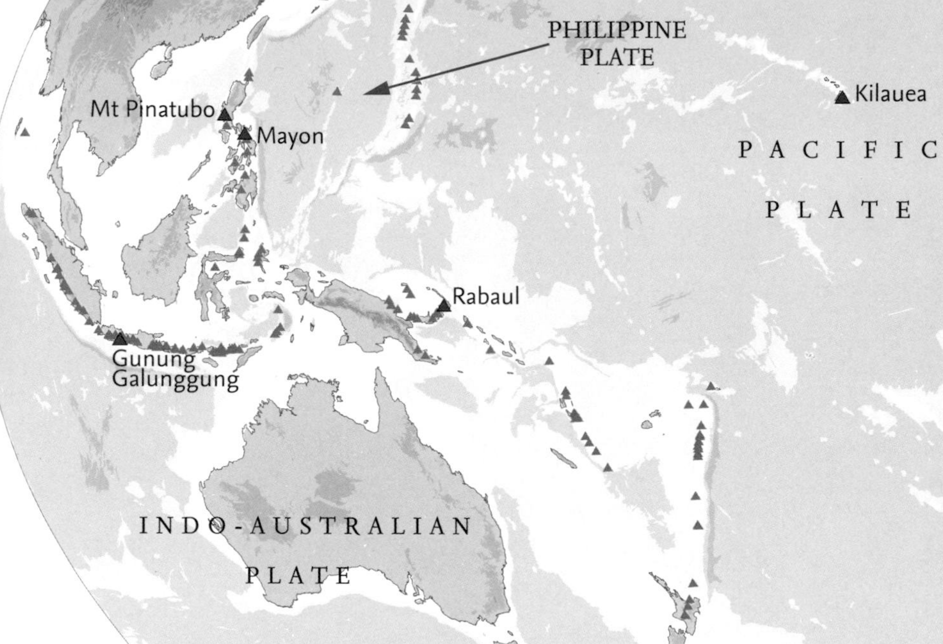

Major volcanic eruptions 1980–2011

Date	Volcano	Country
1980	Mt St Helens	USA
1982	El Chichónal	Mexico
1982	Gunung Galunggung	Indonesia
1983	Kilauea	Hawaii, USA
1983	Ō-yama	Japan
1985	Nevado del Ruiz	Colombia
1991	Mt Pinatubo	Philippines
1991	Unzen-dake	Japan
1993	Mayon	Philippines
1993	Volcán Galeras	Colombia
1994	Volcán Llaima	Chile
1994	Rabaul	Papua New Guinea
1997	Soufrière Hills	Montserrat
2000	Hekla	Iceland
2001	Mt Etna	Italy
2002	Nyiragongo	Dem. Rep. of the Congo
2010	Eyjafjallajökull	Iceland

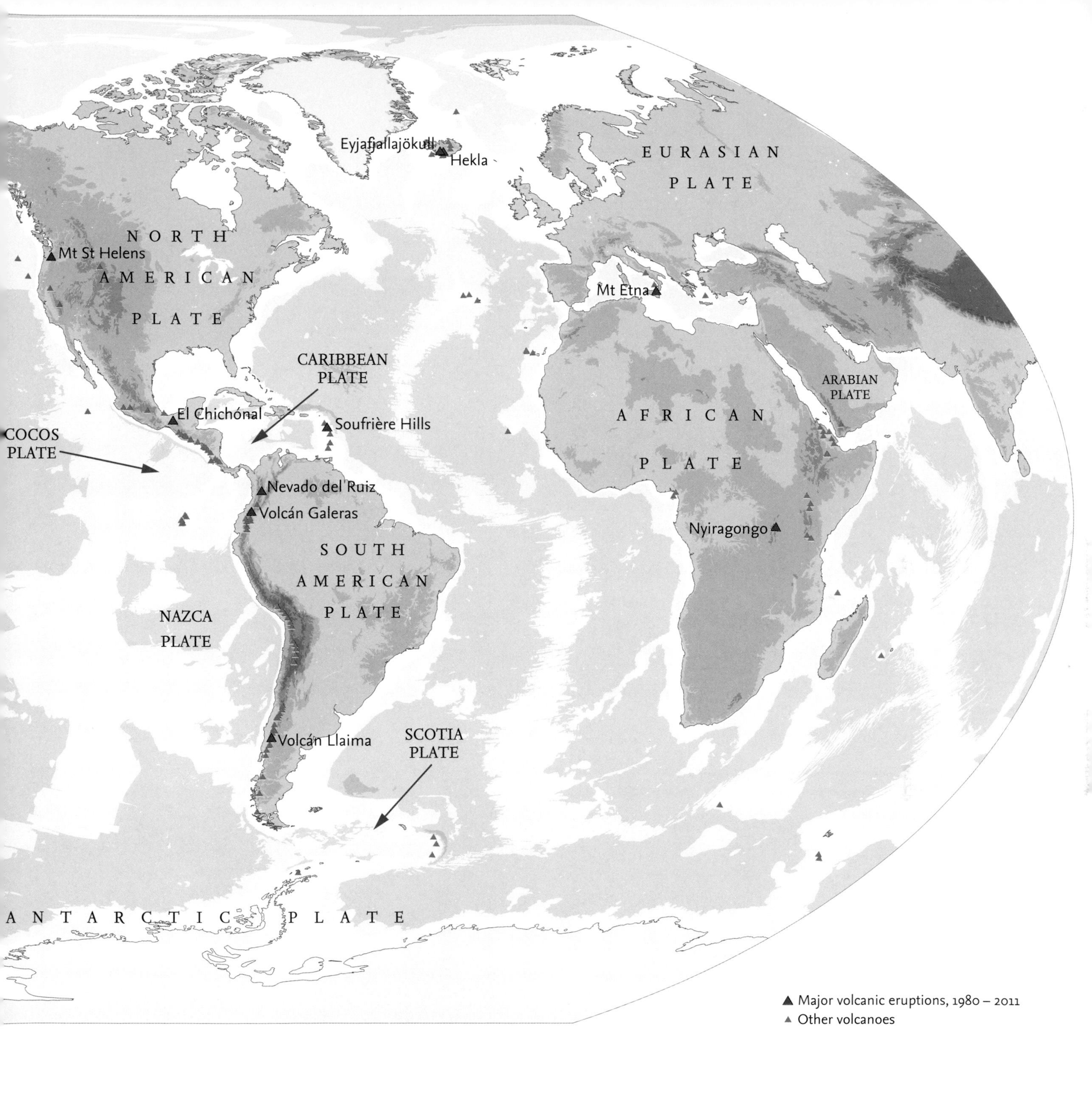

Beneath the solid rock of the Earth's crust lies the mantle, made from dense, hot rock which can flow very slowly. There are certain places where, from time to time, material from the mantle rises to the surface. As it rises the pressure compressing it relaxes and the material expands. Certain of its ingredients vaporize and may explode. Hot gas, molten rock, and blocks of solid rock rise to the surface and flow into the air or ocean. The hot material that is held under pressure just below the surface is called magma. When magma pours across the surface it is called lava. That is a volcanic eruption.

Volcanoes are found where the Earth's crust is thin and where there are fissures at the boundaries between the vast slabs, or plates, of rock which comprise the Earth's crust. They are particularly common around the shores of the Pacific Ocean, where active volcanoes form what is often called a 'ring of fire'.

Principal volcano types

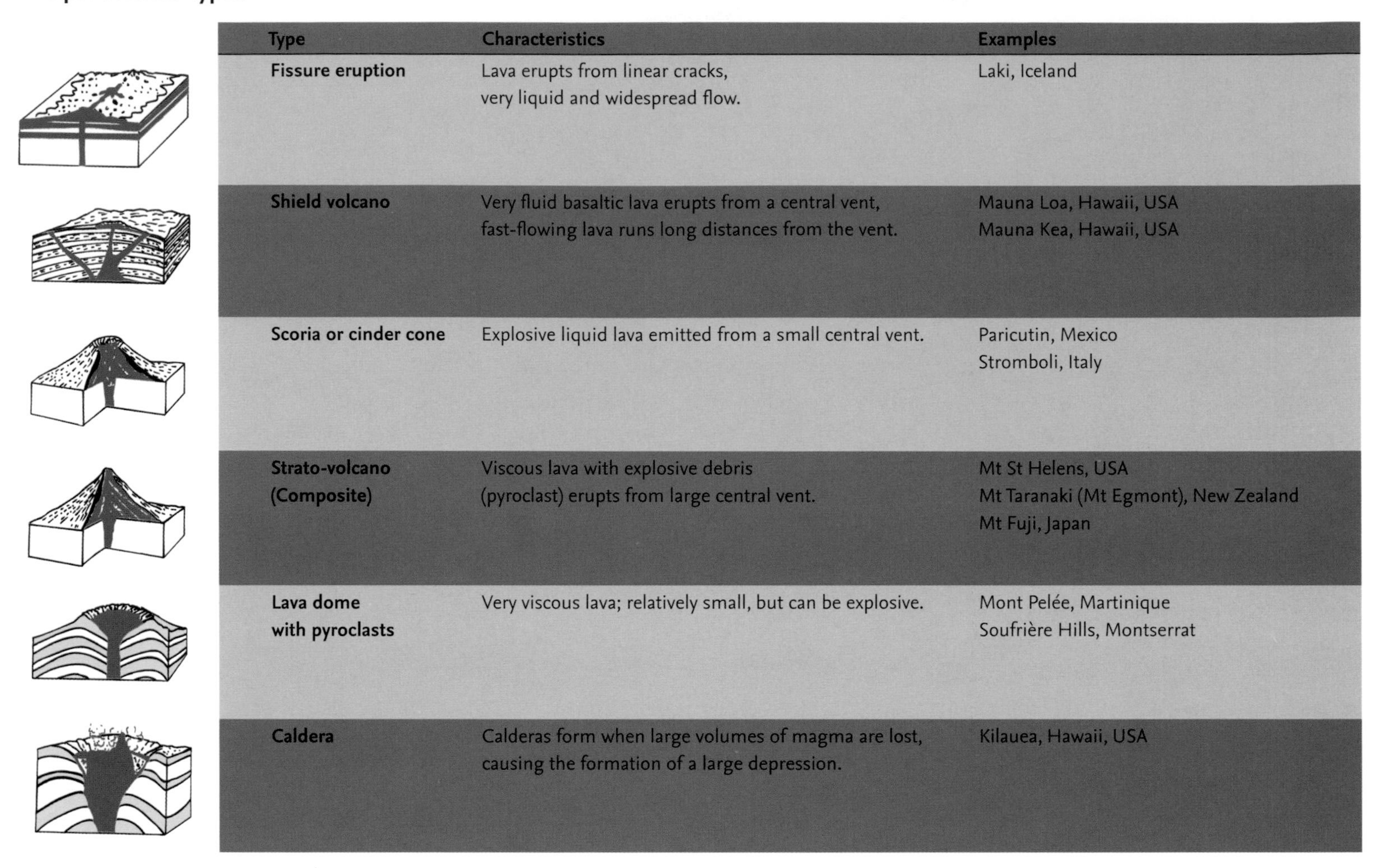

Type	Characteristics	Examples
Fissure eruption	Lava erupts from linear cracks, very liquid and widespread flow.	Laki, Iceland
Shield volcano	Very fluid basaltic lava erupts from a central vent, fast-flowing lava runs long distances from the vent.	Mauna Loa, Hawaii, USA Mauna Kea, Hawaii, USA
Scoria or cinder cone	Explosive liquid lava emitted from a small central vent.	Paricutin, Mexico Stromboli, Italy
Strato-volcano (Composite)	Viscous lava with explosive debris (pyroclast) erupts from large central vent.	Mt St Helens, USA Mt Taranaki (Mt Egmont), New Zealand Mt Fuji, Japan
Lava dome with pyroclasts	Very viscous lava; relatively small, but can be explosive.	Mont Pelée, Martinique Soufrière Hills, Montserrat
Caldera	Calderas form when large volumes of magma are lost, causing the formation of a large depression.	Kilauea, Hawaii, USA

Volcanoes differ in shape and formation – as seen in the table of volcano types above – and in the way they erupt, due to variations in the composition of the magma. Eruption types include, Strombolian, Vulcanian, Hawaiian, Peléean, Surtseyan and Plinian.

Some volcanoes do not grow into mountains at all. Fissure volcanoes, such as Laki, Iceland, erupt along cracks in the crust and their lava flows to the sides, producing a long, low mound. Paricutin, Mexico, and Stromboli, Italy, are scoria cones and the eruptions producing them are Strombolian, named after the Italian volcano. A scoria or cinder cone can be up to 300 m (984 feet) tall. Eruptions are frequent and involve many small explosions caused by the expansion of gases held inside very viscous lava. The lava is thrown upwards and falls back to build the cone. Not far from Stromboli there are volcanoes which produce Vulcanian eruptions. These explode with no release of magma, shattering the overlying solidified lava and hurling it high into the air, mixed with ash and gas. Mt St Helens, Mt Taranaki (Mt Egmont) in New Zealand, and Mt Fuji are strato-volcanoes, built up from layers of solidified lava alternating with layers of ash and loose rock. When they erupt they often do so explosively from a single vent. Shield volcanoes are broad cones with gently sloping sides. They are made from very liquid lava and their eruptions produce spectacular fire fountains. Many Hawaiian volcanoes are of this type.

Montserrat

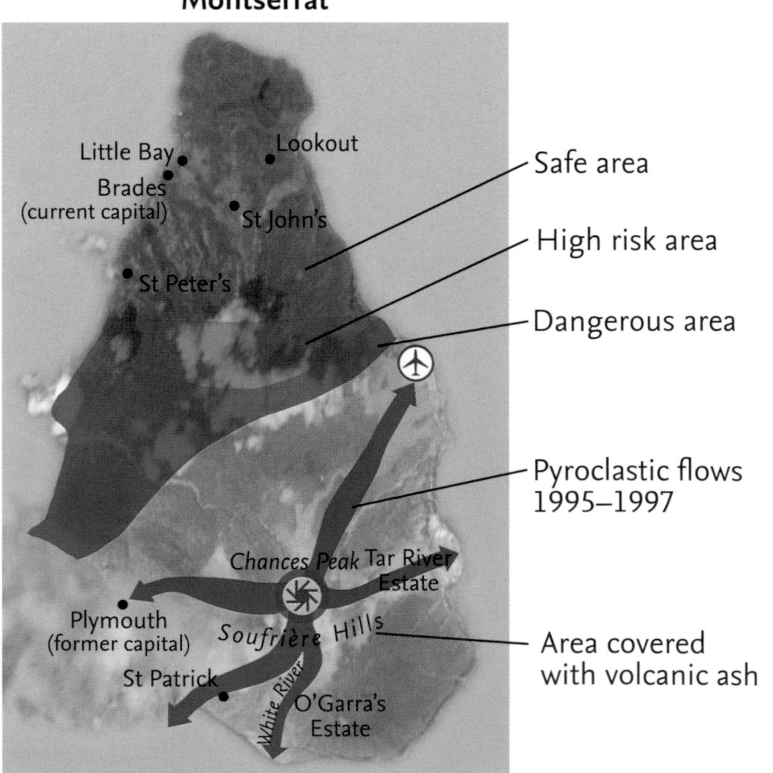

Extent of Eyjafjallajökull ash cloud

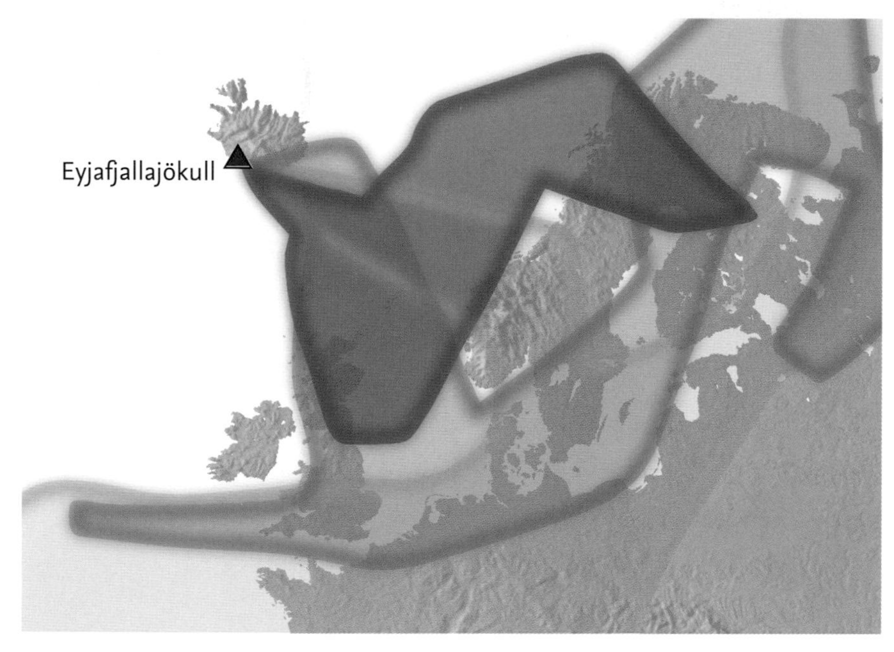

Extent of ash cloud

- 15 April 06:00 GMT
- 16 April 06:00 GMT
- 17 April 00:00 GMT

Eyjafjallajökull in the southwest of Iceland is classed a strato-volcano. Fissures began to open on 20 March 2010 after weeks of earthquakes lifted the land surface in the local area by 40 mm (2 inches). Eruptions through the glacier caused meltwater in the rivers to flood, resulting in the evacuation of 800 people. Under a month later, major eruptions caused volcanic ash to be lifted into the atmosphere, several kilometres into the air, where it began to drift slowly southeast towards Europe. The abrasive nature of the ash caused air traffic to be suspended for at least six days across northwestern Europe and included some trans-Atlantic flights. It is estimated that around 100 000 flights were cancelled.

Montserrat eruption chronology

1992 1993 1994 1995 1996 1997 1998

January 1992 Start of earthquake swarms in southern Montserrat.

18 July 1995 Initial steam and ash venting from the northwest crater.

21 August 1995 First large eruption blankets Plymouth in a thick ash cloud and causes darkness.

3 April 1996 First pyroclastic flow, which travelled to the road which crosses the Tar River valley.

12 May 1996 Pyroclastic flows reached the sea for the first time.

17–18 September 1996 First magmatic explosion with a major ash plume. 600 000 tonnes of ash was deposited in southern Montserrat.

1–11 April 1997 Major pyroclastic flows down White River nearly reaching the sea at O'Garra's.

25 June 1997 Major pyroclastic flows in Mosquito Ghaut reaching to within 50 m of the airport.

3 August 1997 Major pyroclastic flows into Plymouth.

21 September 1997 Major dome collapse to northeast of the volcano, destroying airport terminal building and entering the sea at various points along the coast.

Peléean eruptions, named after Mont Pelée in Martinique, which erupted violently in 1902 and more recently in 1995–6, are often preceded by the accumulation of magma pushing the overlying rock upwards, forming a lava dome which then bursts. Such eruptions release huge clouds of extremely hot ash and small rocks in a pyroclastic flow which moves at great speed, burning and burying everything in its path. Volcanic ash consists of finely powdered rock which solidifies on contact with moisture. Inhaling it can be fatal. The eruption on Montserrat detailed on this page was also of this type.

THE ERUPTION OF EYJAFJALLAJÖKULL, ICELAND 16 APRIL 2010

After lying dormant for nearly two centuries, Iceland's Eyjafjallajökull volcano started erupting on 20 March 2010. The eruption was small and lasted only a few days. However, the second eruption which began on 14 April 2010 was far more dramatic, and the resulting ash plumes caused huge disruption to air travel across Europe. Indeed, most European airspace was closed from 15–21 April, and over 100 000 flights were cancelled in total at a great financial cost to the European economy. The ash clouds

reached heights of 10 km (6 miles), and spread in a southeastern direction to cover vast areas of Europe. The reason behind the ash plumes being so dramatic lies in the fact that the volcano is located under an icecap; during the eruption vast quantities of melting water ran into the crater and merged with the hot lava, creating an explosive mix that produced the enormous plumes of ash.

A TOWN ENGULFED… PLYMOUTH, MONTSERRAT 1997

Plymouth, the former capital of the British territory of Montserrat in the Caribbean, was buried in volcanic ash as a result of eruptions of the Soufrière Hills volcano several times between 1995 and 1997. The town, which was the island's largest settlement, had to be abandoned and many of the residents moved away from the island and have not returned. The capital moved to the northern settlement of Brades temporarily.

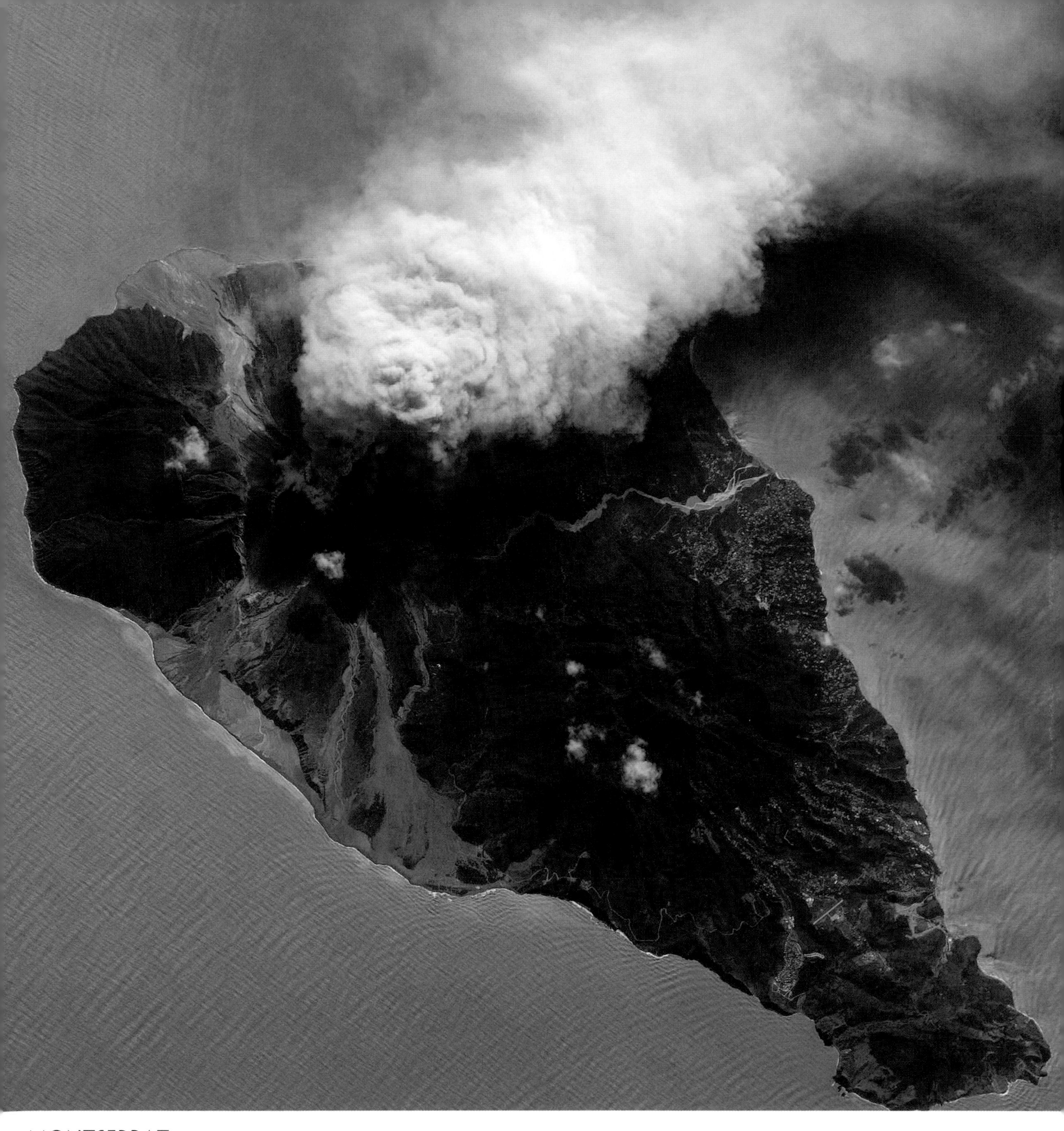

MONTSERRAT 2009

In October 2009 the island experienced earthquakes and ash clouds appeared. These plumes can deposit ash at great distances from the island. Further activity in February 2010 deposited ash on Guadeloupe and Antigua. Although parts of the island remain inhabited and many of the services lost when Plymouth was destroyed have been moved elsewhere, this image shows how vulnerable the remaining population is to another large eruption. Brades is still being used as the capital.

BIRTH OF AN ISLAND, ZUBAIR ISLANDS, RED SEA 24 OCTOBER 2007 AND 23 DECEMBER 2011

In mid-December 2011, fishermen observed the beginnings of an undersea eruption where lava fountains reached heights of 30 m (98 feet) above sea level. The first satellite image from 24 October 2007 shows the area before the eruption occurred. The second satellite image from 23 December 2011 shows the eruption in full swing, with plumes of volcanic ash and water vapour rising out of the sea. By this point the lava mass had broken through the sea's surface, and the new island started to take shape as the erupting lava was cooled by sea water and started to solidify into rock.

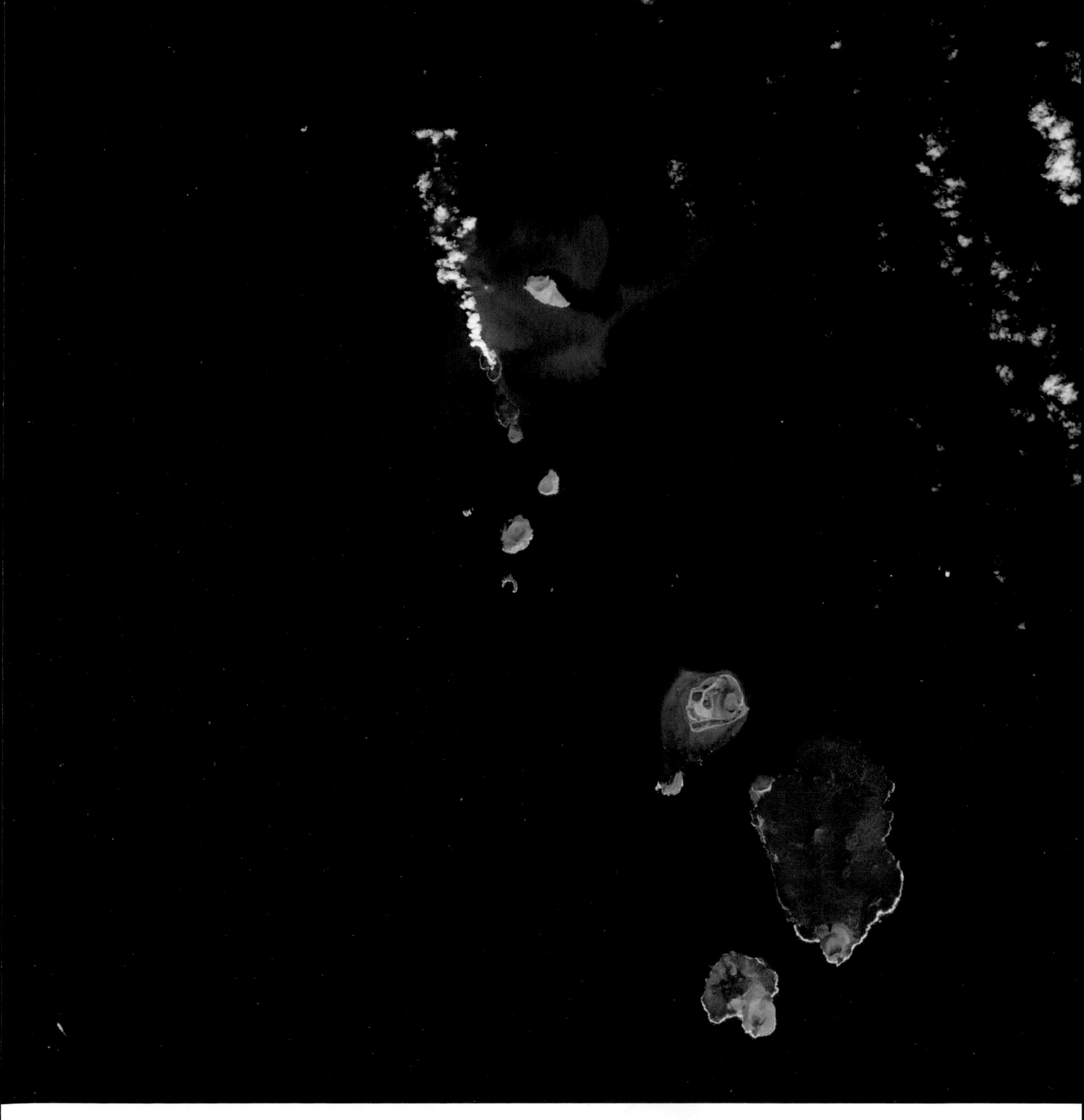

NEW ISLAND, ZUBAIR ISLANDS, RED SEA 15 JANUARY 2012

One month later, a satellite image shows that volcanic activity had ceased, and a new island was visible between Haycock and Rugged islands. The new island measures 530 by 710 m (1740 by 2330 feet) across, and is part of the Zubair Islands group, which lies 60 km (37 miles) off the coast of Yemen.

The islands rise above the sea surface in a northwest to southeast line. They are part of a large underwater shield volcano, which is located on the Red Sea Rift where the African and Arabian tectonic plates are gradually pulling apart from one another.

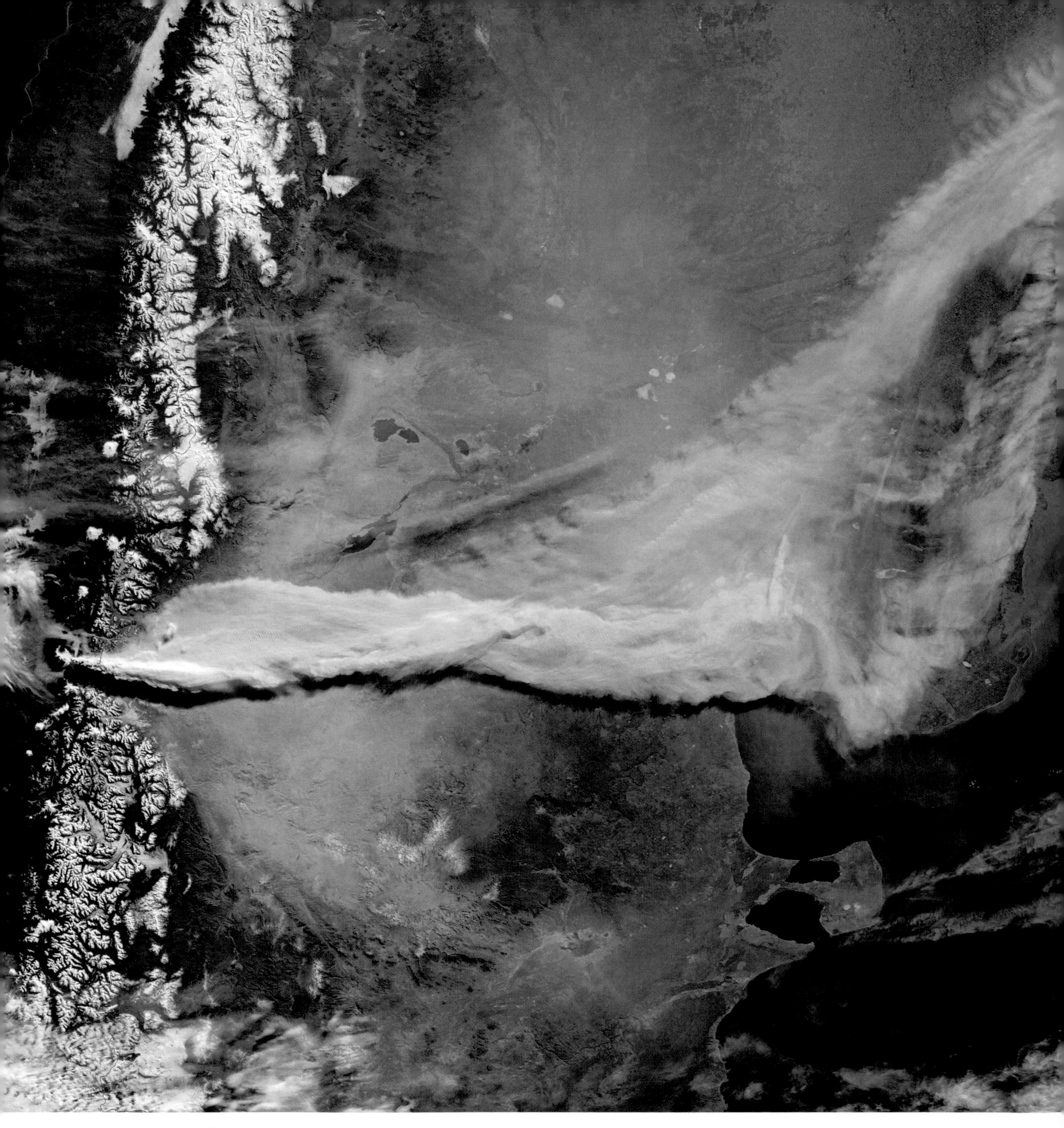

PUYEHUE-CORDÓN ERUPTION 13 JUNE 2011

On 4 June 2011, after five decades of inactivity, an eruption began in the Puyehue-Cordón volcanic chain in south-central Chile. This satellite image shows the large volcanic ash plume that travelled east and then northeast over Argentina, causing heavy ash falls in Argentinian towns up to 100 km (60 miles) away. As the volcanic ash circled the globe it caused widespread air-travel disruption, including countries as far away as Australia and New Zealand.

PUYEHUE-CORDÓN ERUPTION 5 JUNE 2011

During the eruption, plumes of ash reached heights of 10 km (6 miles). As a precaution the government ordered the evacuation of thousands of residents from towns and villages around the volcanic chain. On the Argentinian border town of Bariloc, ash fell to a depth of 300 mm (12 inches).

TSUNAMI – large, often destructive, sea waves produced by submarine earthquakes, subsidence, or volcanic eruptions

Tsunami hitting the coast of Minamisoma, Fukushima prefecture, Japan, 11 March 2011

TSUNAMIS

Cross-section of a tsunami

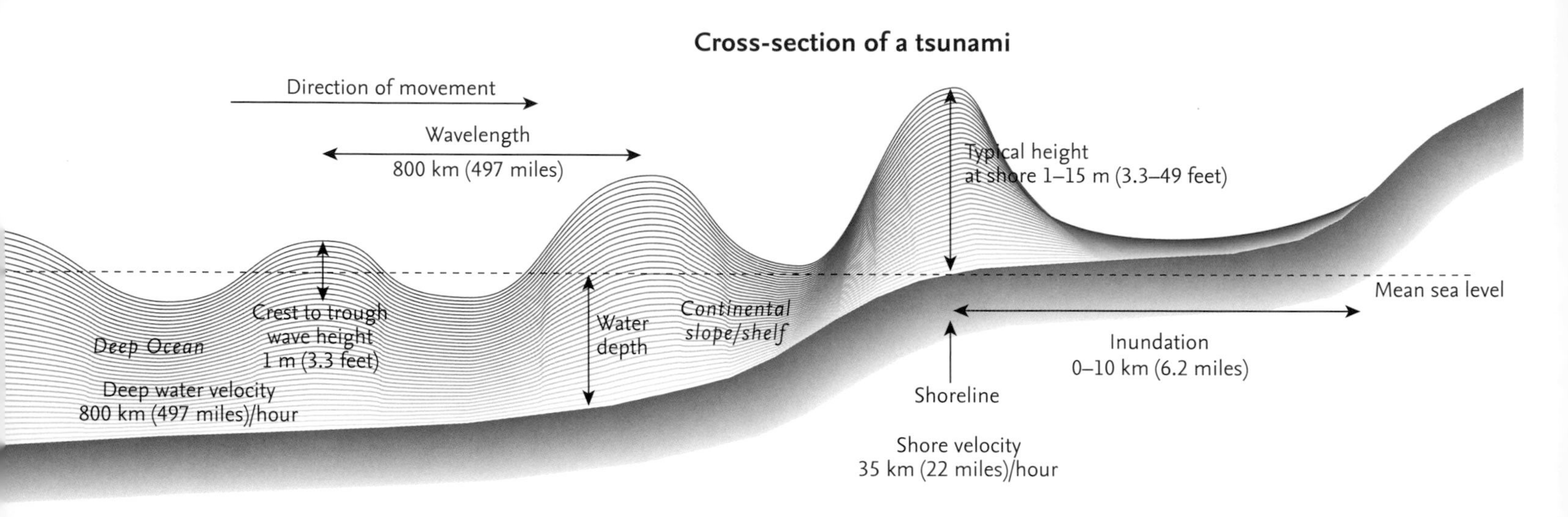

Tsunamis in recent history

Date	Location	Extent	Caused by	Deaths
8 November 1929	Grand Banks of Newfoundland, Canada	Atlantic Ocean	Earthquake of magnitude 7.2 and submarine landslide	29
1 April 1946	Aleutian Islands, Alaska, USA	Pacific Ocean	Earthquake of magnitude 7.8	165
9 March 1957	Aleutian Islands, Alaska, USA	Pacific Ocean	Earthquake of magnitude 8.2	-
4 November 1957	Kamchatka Peninsula, Russian Federation	Pacific Ocean	Earthquake of magnitude 8.3	-
22 May 1960	South Central Chile	Pacific Ocean	Earthquake of magnitude 8.6	122
28 March 1964	Prince William Sound, Alaska, USA	Pacific Ocean	Earthquake of magnitude 8.6	-
29 November 1975	Hawaii, Pacific Ocean	Local	Earthquake of magnitude 7.2	2
17 July 1998	Papua New Guinea	Local	Earthquake of magnitude 7.0 and submarine landslide	2 200
26 December 2004	off Sumatra, Indonesia	Indian Ocean	Earthquake of magnitude 9.0	226 408
11 March 2011	off Honshū, Japan	Pacific Ocean	Earthquake of magnitude 9.0	15 800

Terrifying and destructive as earthquakes can be, sometimes they give rise to another phenomenon which can cause even more destruction and loss of life – the tsunami. When an earthquake occurs offshore, it may cause a sudden change in the shape of the ocean floor, as a result of vertical fault movement or submarine landslides. This causes a massive displacement of water, which in turn produces a powerful wave or series of waves capable of travelling over huge distances. Thankfully, although earthquakes are frequent occurrences, tsunamis resulting from them are relatively rare and it is even rarer for them to cause significant loss of life.

Tsunamis can travel at great speed in the open ocean. The tsunami of 26 December 2004, saw the wave travelling at over 800 km (500 miles) per hour across the Indian Ocean. The Japan tsunami of 11 March 2011 took only twenty hours to cross the Pacific Ocean to reach the South American coast. In deep ocean water, the wave itself may seem small and insignificant, with heights of only 1 m (3 feet) greater than normal. However, as such waves reach shallower water their speed decreases but their height increases dramatically to create highly destructive waves which can be over 15 m (50 feet). The local coastal topography and shape of the sea bed influences the final effect of a tsunami, but the forces involved are enormous. The force of the water caused the wave to travel up to 10 km (6 miles) inland across the low-lying coastal plains of Japan which resulted in millions of tonnes of debris as villages were wiped away and left over 20 500 people dead or missing.

2004 Indian Ocean tsunami travel times (hours)

Earthquakes off Japan 2011

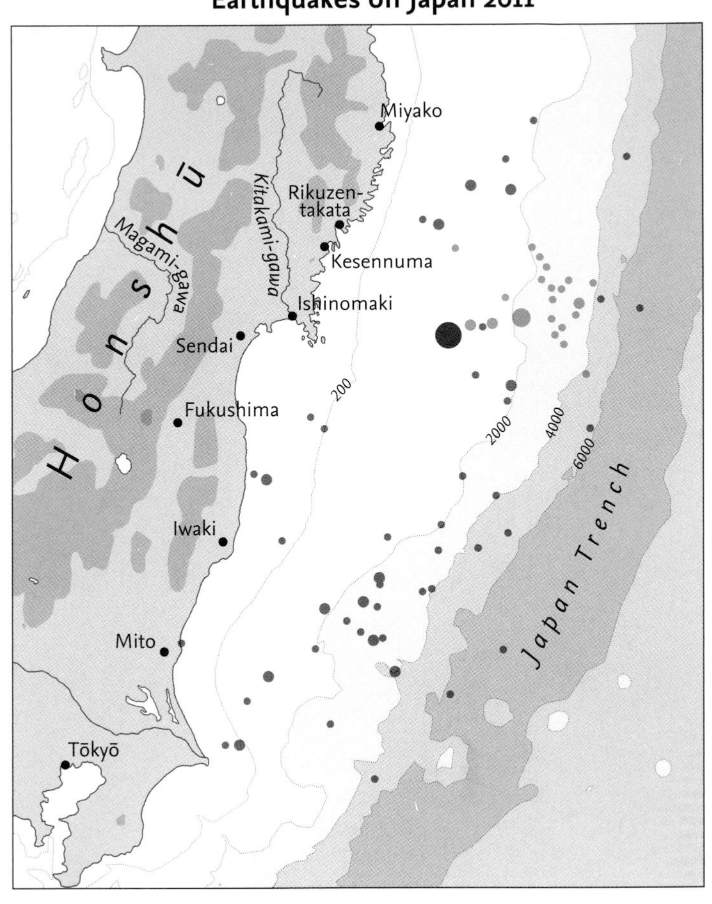

2011 Pacific Ocean tsunami travel times (hours)

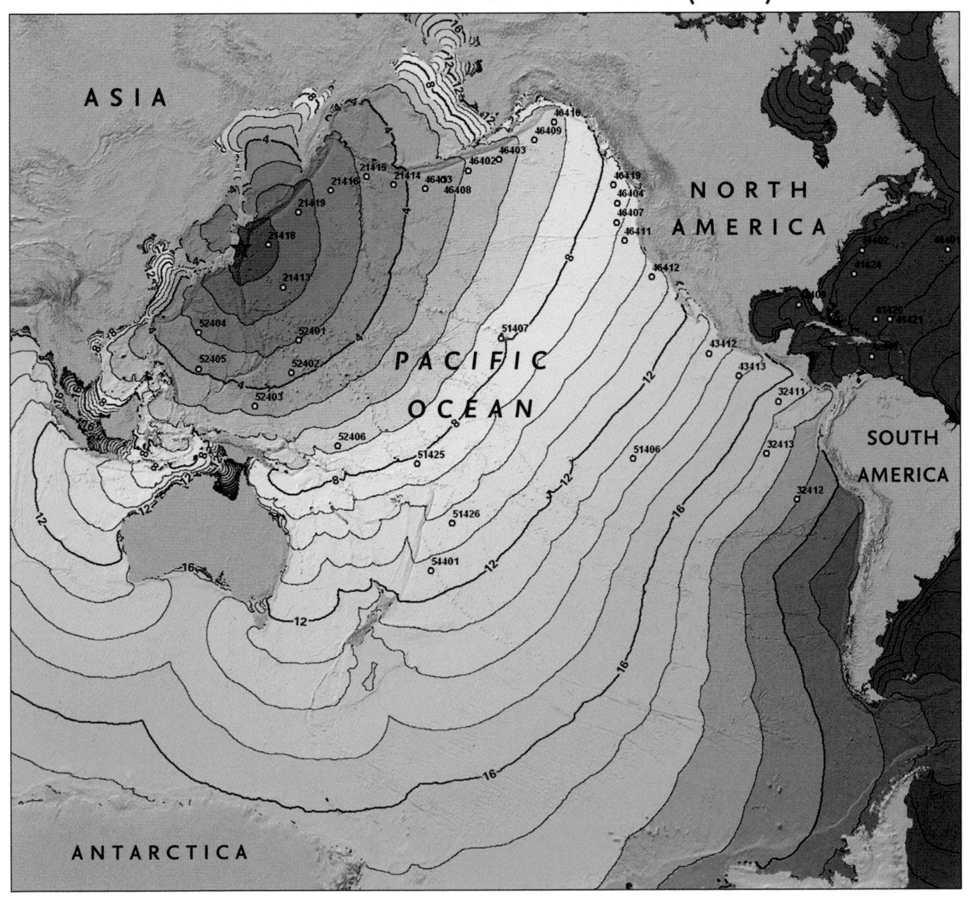

Shocks from 9 March until main shock: 5–5.9, 6–6.9, 7–7.9
Main shock on 11 March: 8–8.9
After shocks: 5–5.9, 6–6.9, 7–7.9
earthquake magnitude

The most dangerous types of boundary between tectonic plates are those along subduction zones, where one plate is forced under another. Great pressure builds up over centuries as the plates converge. This pressure is released by the overlying plate slipping back into position when the rocks can no longer bear the pressure. The east coast of Japan took the full force of the tsunami which originated 130 km (80 miles) east of Sendai but 2m (6 feet) waves were measured across the Pacific Ocean in California. The Sumatra tsunami affected a greater number of countries on the fringes of the smaller Indian Ocean. The after effects of the tsunami will be felt for years. Not only in the reconstruction on land but it is estimated there is up to two million tonnes of debris floating in the Pacific Ocean, some of which is expected to cross the ocean and reach the American coast. The first significant piece was an empty ship which appeared off the coast of British Columbia, Canada. Such items could continue into 2014.

RIKUZEN-TAKATA, JAPAN JULY 2010 BEFORE…

Rikuzen-takata is situated on the coast in the Tōhoku region of Japan. This is a relatively young city, being founded in 1955. In October 2008 it had an approximate population of 24 000 people. Further along the coast was an area designated a 'Place of Scenic Beauty' with about 70 000 pine trees.

RIKUZEN-TAKATA, JAPAN MARCH 2011 AFTER…

Following the earthquake and subsequent tsunami in March 2011, the city was destroyed. Although the city was prepared for both earthquakes and tsunamis, the events on 11 March far exceeded anything expected. The tsunami shelters were built to withstand waves of 3–4 m (9–13 feet) in height, but the 2011 tsunami wave was 13 m (42 feet) in height. Approximately 80 per cent of all the houses in the city were swept away. Only one pine tree of the 70 000 remained standing and it is now under threat from the salts in the soil.

ISHINOMAKI, MIYAGI PREFECTURE, JAPAN 15 MARCH 2011 AND 13 JANUARY 2012

The low-lying and heavily-urbanized port city of Ishinomaki was one of the hardest hit by the 2011 Tōhoku earthquake and tsunami. The waves reached heights of up to 40 m (131 feet), and travelled up to 10 km (6 miles) inland from the coast. In Ishinomaki over 28 000 houses were destroyed. The top image shows a bridge covered in wreckage just days after the tsunami. Almost one year on and the bridge is functional again, the flood waters have receded, debris has been removed and there is ongoing work to restore and rebuild the city's infrastructure and buildings.

KESENNUMA, MIYAGI PREFECTURE, JAPAN 16 MARCH 2011 AND 14 JANUARY 2012

Large sections of the port city of Kesennuma in Miyagi Prefecture were destroyed by the 2011 Tōhoku earthquake and tsunami. After the event, spilled fuel from the city's heavily-damaged fishing fleet caught fire and burned for four days. The first photo from 16 March 2011 shows part of the city just days after it had been engulfed by the tsunami waves; flood water is evident, along with piles of debris, fishing vessels which were washed inland, and empty spaces where buildings and houses used to stand. Nearly one year on and the flood waters have receded and debris has been cleared.

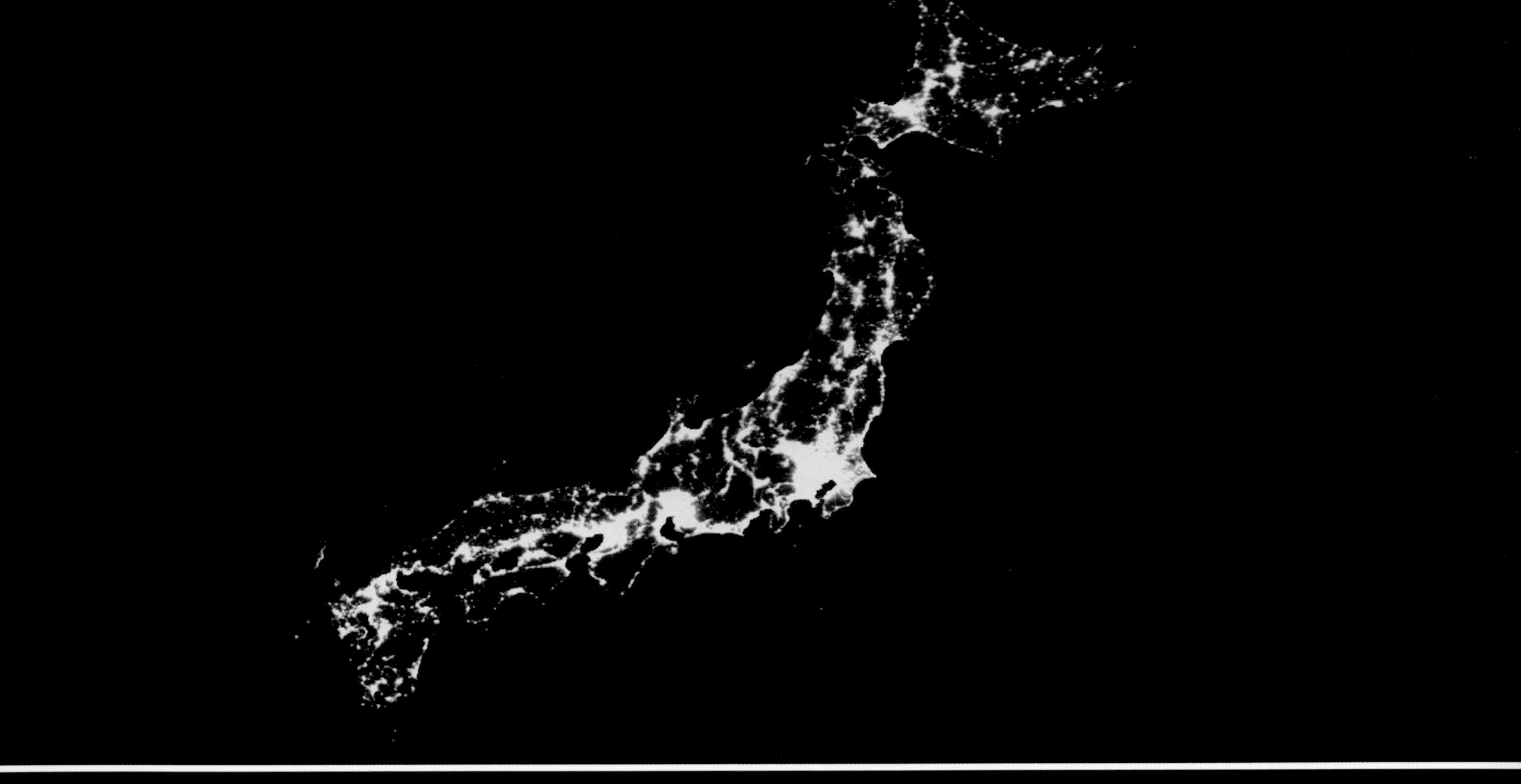

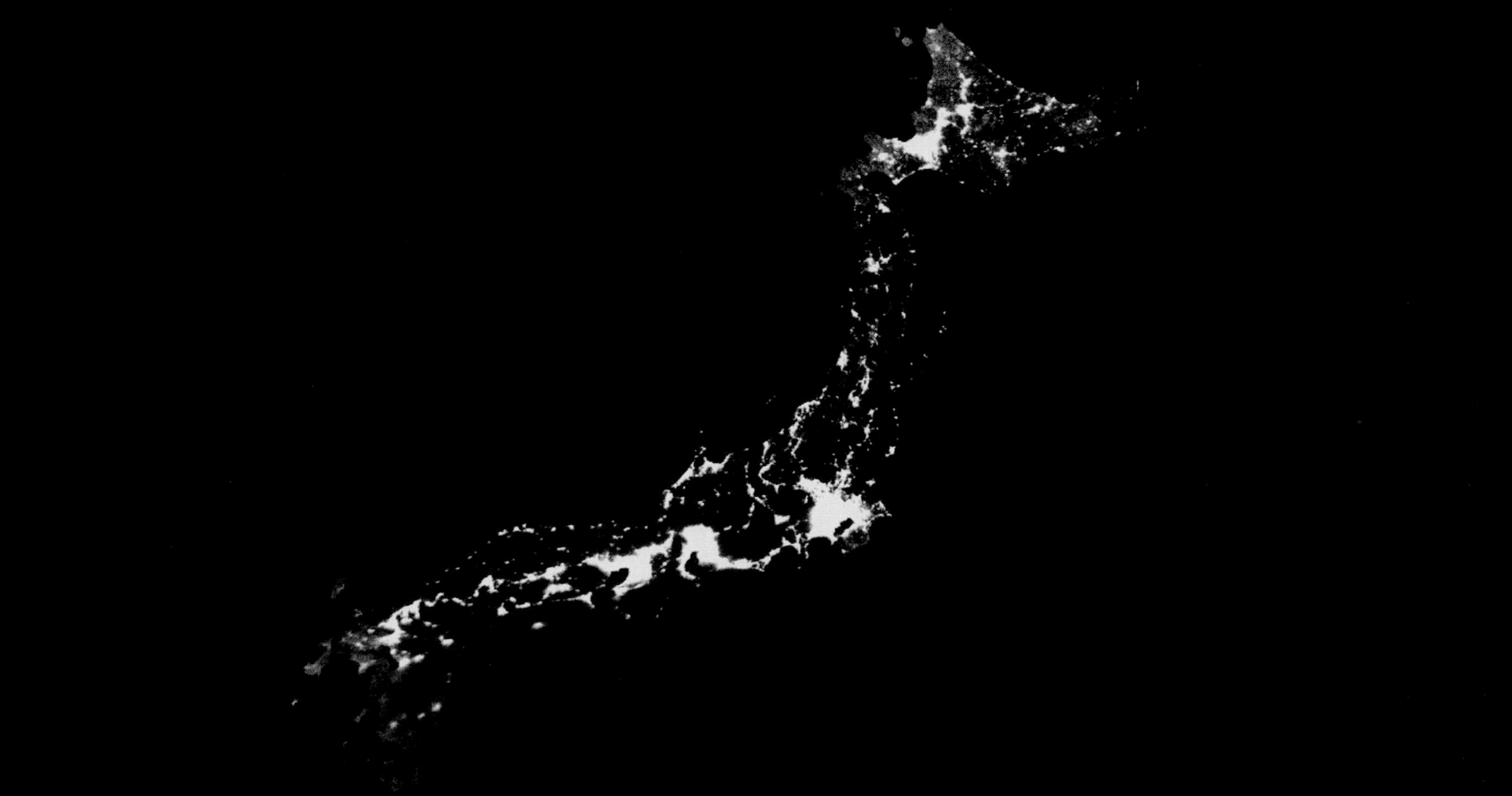

JAPAN AT NIGHT 2010 AND 12 MARCH 2011

The pair of satellite images above demonstrate the catastrophic effect of the 2011 Tōhoku earthquake and tsunami. The first satellite image shows the country at night a year before the tsunami event. The most densely populated areas are brighter and the outline of the country can clearly be made out. In contrast, in the second satellite image the impact of the tsunami is evident as the northeastern coast is almost completely blacked out. Light levels were also much reduced over the rest of Japan as the country's electrical power network struggled to cope.

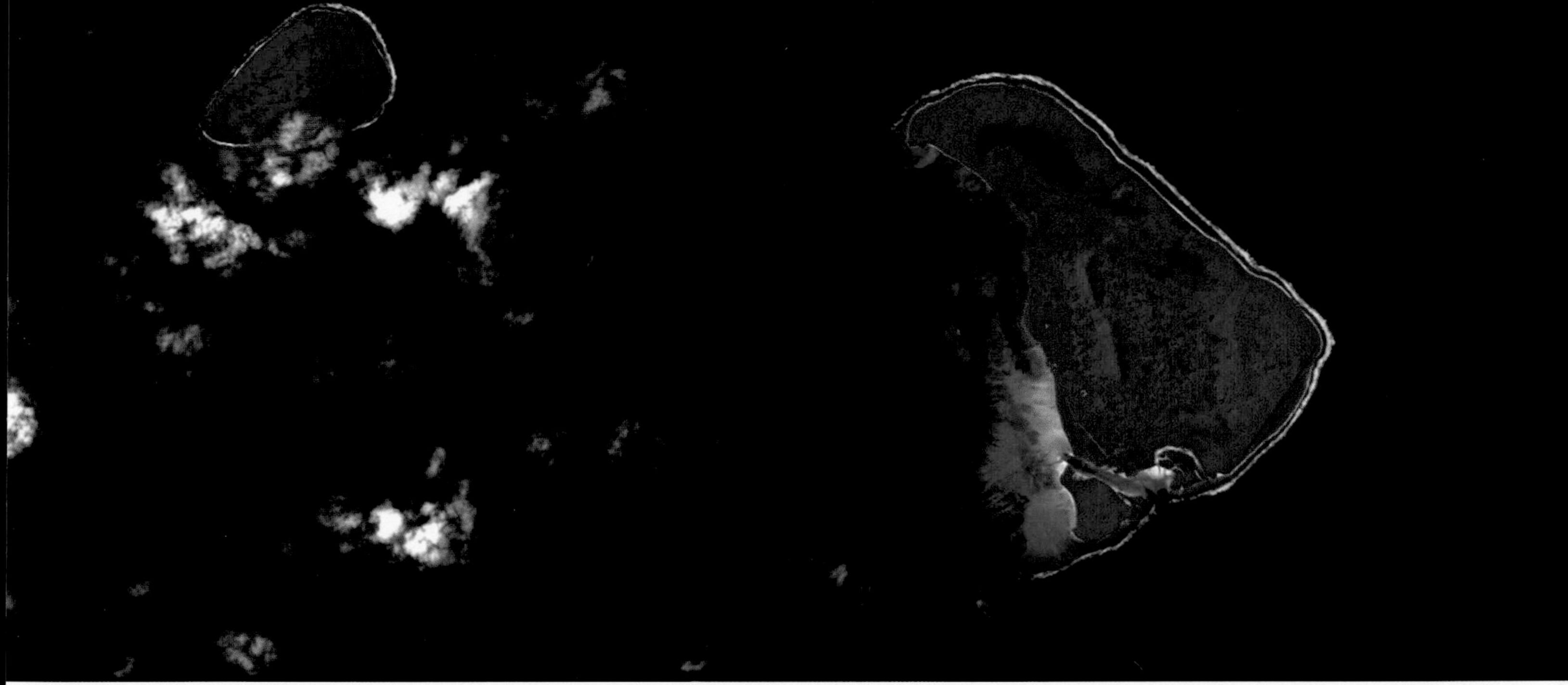

NIUATOPUTAPU, TONGA 25 JULY 2006 AND 19 OCTOBER 2009

False-colour satellite images are useful tools in analysing vegetation change over large areas; vegetation is red in colour, bare ground is pale blue-grey, clouds and breaking waves are white, and water is blue or black. On 29 September 2009, a magnitude 8.1 earthquake located off Samoa generated a tsunami; the 6 m (20 feet) high waves hit Niuatoputapu, damaging two towns and killing nine island residents. In the second image, the dark red area on the northeast side of the island indicates damaged or inundated vegetation and shows how far inland the waves travelled.

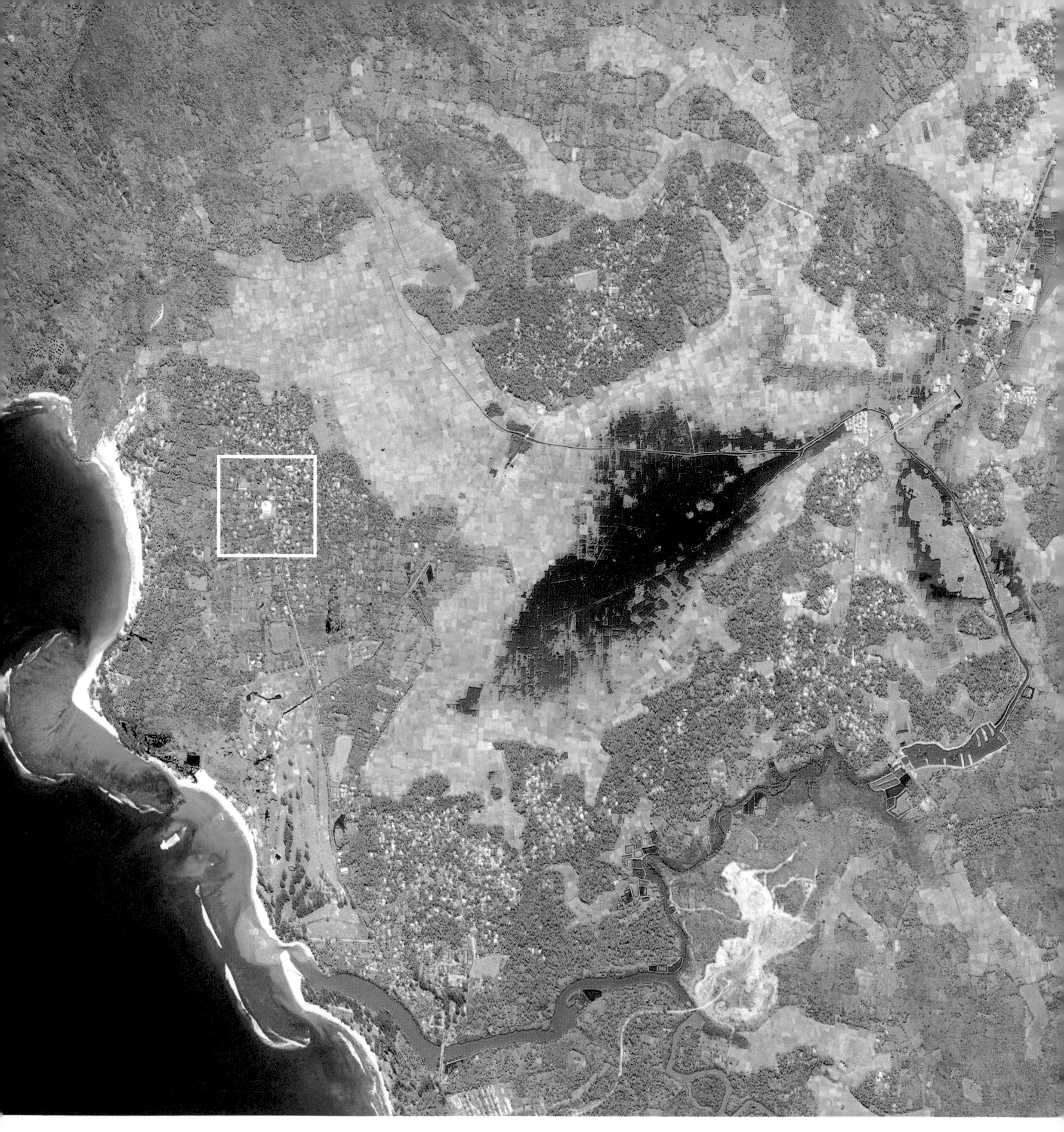

LHOKNGA, ACEH, INDONESIA 10 JANUARY 2003

Captured in January 2003, this satellite image of Lhoknga, near the provincial capital of Banda Aceh, shows a region of lush and well-cultivated land, with woodlands and several villages. The darker area in the centre is water and there are several irrigation canals. The coast has sandy beaches, some with barrier islands or reefs protecting them.

TSUNAMI AFTERMATH... 29 DECEMBER 2004

Three days after the tsunami the extent of the destructive force of the waves can be seen. The coastal area has been stripped bare of vegetation and buildings with only the prominent Rahmatullah Lampuuk Mosque remaining. Inland, the low-lying areas are now swamped with salt water and it is only the slightly higher level of the roads which keeps them visible. In 2012 two major earthquakes in the same area as the 2004 episode, triggered fears of another devasting tsunami. But the resultant wave measured only 80 cm (31 inches) greater than normal.

LANDSLIDES AND AVALANCHES –

movements of large masses of rock, down mountains or cliffs

Landslide destroys a bridge

LANDSLIDES AND AVALANCHES

Deadliest landslides 1900–2011

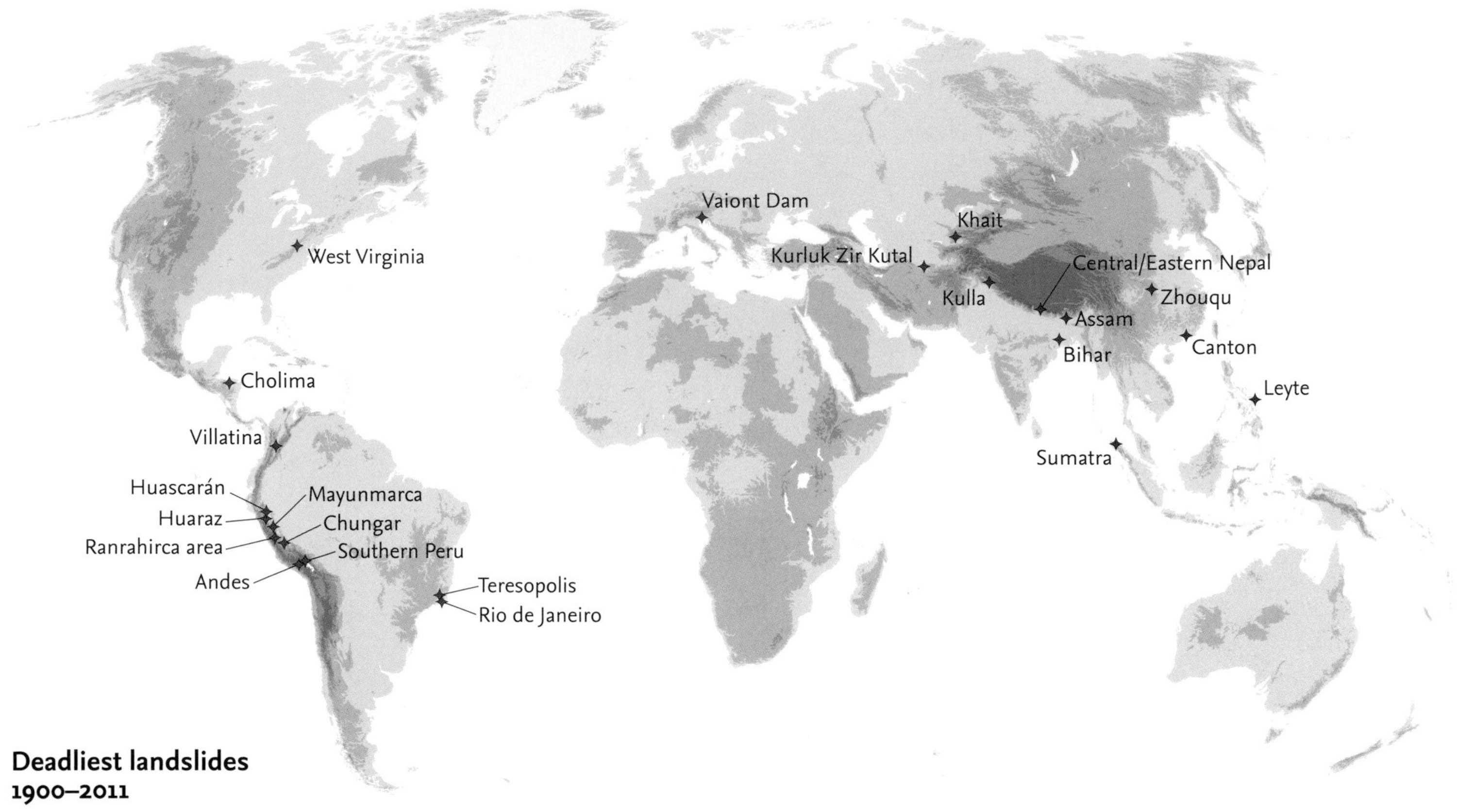

Deadliest landslides 1900–2011

Date	Location	Country	Deaths
1934	Canton	China	500
1941	Huaraz	Peru	5 000
1948	Assam	India	500
1949	Khait	Tajikistan	12 000
1955	Sumatra	Indonesia	405
1962	Ranrahirca area	Peru	2 000
1963	Vaiont Dam	Italy	2 600
1967	Rio de Janeiro	Brazil	436
1968	Bihar, Bengal	India	1 000
1970	Huascarán	Peru	20 000
1971	Chungar	Peru	600
1972	West Virginia	USA	400
1973	Cholima	Honduras	2 800
1973	Andes	Peru	500
1974	Mayunmarca	Peru	310
1987	Villatina, Medellin	Colombia	640
1995	Kulla, Himachal Pradesh	India	400
1995	Kurluk Zir Kutal, Badakhshan Province	Afghanistan	354
1997	Southern Peru	Peru	300
2002	Central/Eastern Nepal	Nepal	472
2006	Leyte	Philippines	139 (980 missing)
2010	Zhouqu, Gansu	China	1 100
2011	Teresopolis	Brazil	440

Steep slopes are often unstable. An earth tremor, heavy rain, or a storm may detach rocks or soil from the underlying solid rock. Gravity will do the rest, sending the loose material downhill as a landslide or, if rain has turned the soil to mud, as a mudslide.

Rock falls, landslides, and mudslides can bury entire villages, with huge loss of life. They can also cause disasters indirectly. In October 1963, ten days of heavy rain destabilized the side of Mount Toc, overlooking the reservoir behind the newly built Vaiont Dam in northern Italy. The surface of the mountain consisted of clay covered by loose rock and as the reservoir filled the rocks slid slowly downhill. Then the pressure of water from the saturated mountainside dislodged the rocks. Late on the evening of 9 October more than 240 million cubic metres (314 million cubic yards) of rock, sliding at more than 110 km (68 miles) per hour, fell into the reservoir. The splash sent huge waves about 300 m (984 feet) up the valley sides and water overflowed the dam wall. A wall of water 70 m (230 feet) high rushed down the valley, inundating four villages and killing more than 2 500 people.

An avalanche is a mass of soil, rock, but more commonly snow, sliding down a slope. There is a risk of avalanches where a thick layer of snow accumulates on a slope of 30–40 degrees. There are two types of avalanche. A point-release avalanche affects only the surface layer and typically is shaped like an inverted V. A slab avalanche is more dangerous. It happens when an entire block of snow up to 800 m (2 625 feet) wide is dislodged and moves at up to 200 km (124 miles) per hour, carrying everything before it. A slab avalanche also pushes air ahead of the snow, producing an avalanche wind blowing at up to 300 km (186.4 miles) per hour.

Anatomy of an avalanche

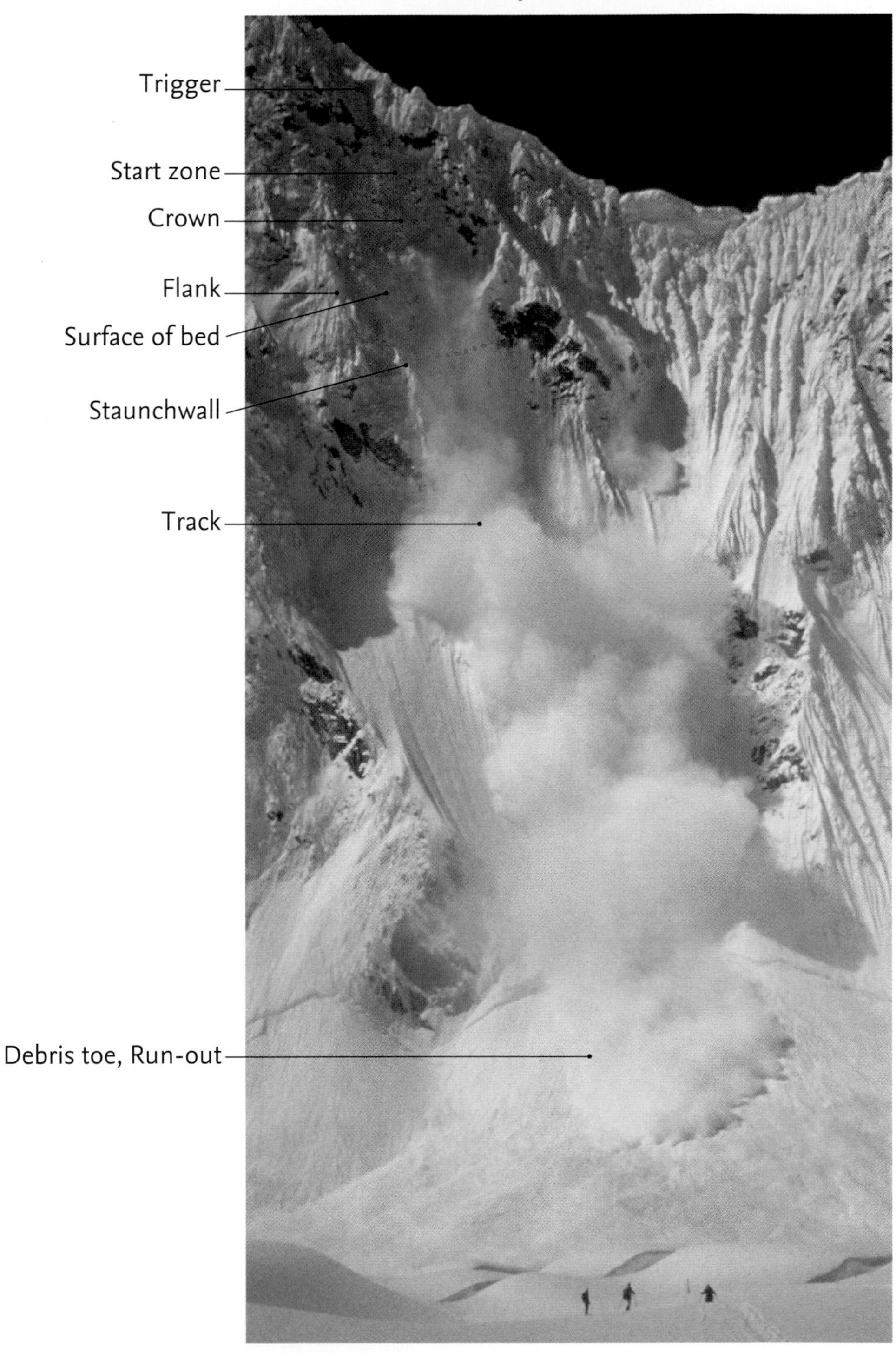

Landslide types

Type	Rate of movement	Material	Occurs along/in	Affected by
flow	rapid	loose soil, rock, water	steep slopes	surface water flow
fall	abrupt	rocks and boulders become detached	fractures, joints, bedding planes	gravity, weathering, water
topple	abrupt	forward rotation of units	pivotal point	gravity, force, water
slumps and slides	slow	soft rocks and sediments	planes of weakness	gravity, water,
creep	very slow and steady	soil or rock	areas of shear stress	water, temperature, material strength

HOLBECK HALL HOTEL, SCARBOROUGH, ENGLAND 4 JUNE 1993 AND FOUR DAYS LATER

On 3 June 1993 Holbeck Hall Hotel was a four-star establishment standing about 65 m (213 feet) above the sea on the South Cliff, Scarborough on the East Yorkshire coast, UK. It looked out over an expanse of lawn to panoramic views of the North Sea. But by the 6 June, as a result of a massive landslide which took place in four stages, the lawn had disappeared and the ground had collapsed under the whole of the seaward wing of the hotel.

ON THE EDGE... 5 JUNE 1993

The rest of the hotel was unsafe and had to be demolished, but as the slip had been progressive everyone was evacuated without injury. In 1997 the hotel owners took Scarborough Borough council to court looking for damages, alleging that they had not taken all measures to prevent the landslip, but the claim was rejected. Further erosion has completely wiped the area of any evidence of the hotel.

NEELUM RIVER, PAKISTAN 15 SEPTEMBER 2002 AND 9 OCTOBER 2005

These satellite images show the Neelum river at Makhri just north of Muzaffarabad, before and after the magnitude 7.6 earthquake which struck northern Pakistan on 8 October 2005. Major landslides have blocked the river's usual course, forcing it to change direction. Its water is brown with sediment from many more landslides upriver.

SICHUAN LANDSLIDES, CHINA 19 FEBRUARY 2003 AND 23 MAY 2008

The earthquake which struck Sichuan on 12 May 2008 created devastation in the wider landscape as well as in towns and cities. Many landslides were triggered which caused problems with rescue efforts. This area is around 150 km (90 miles) from the epicentre. The false colour images show vegetation in red and the later image has many grey patches where bare ground has been exposed by the many landslides heading downwards into the rivers.

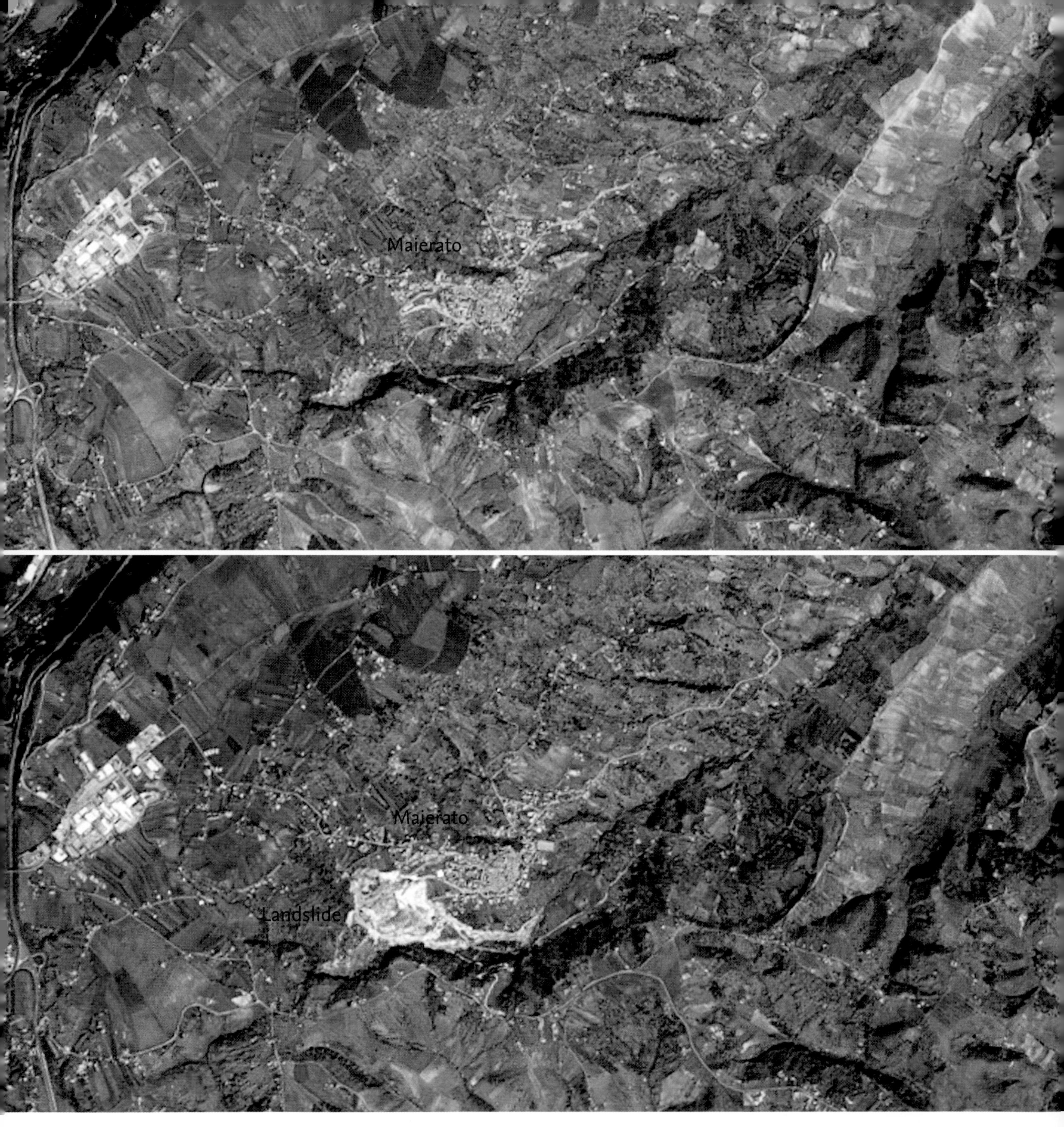

LANDSLIDE IN MAIERATO, ITALY 13 MARCH 2003 AND 14 MARCH 2010

The land around Maierato in Calabria, southern Italy, is steep with a good cover of vegetation. On 15 February 2010 there was heavy rain in the area and it is thought that this contributed to the landslide which happened just outside the town. The town can be seen in the centre of the image. About 200 residents had to be evacuated.

LANDSLIDE IN MAIERATO, ITALY 16 FEBRUARY 2010

The slope which failed had been under stress for some time so the closest residents had been safely evacuated before the slip. There were around 100 other small landslides in the general area at the same time although these do not seem to have affected populated areas except in San Fratello in Sicily.

MOUNT COOK, NEW ZEALAND BEFORE...

Mount Cook, or Aoraki, is an imposing sight. In this image it is 3764 m (12 349 feet) high, but the steep slopes are unstable with hanging glaciers clinging to its slopes. On the night of 14 December 1991, climbing parties heard loud rumbles and saw bright sparks flashing in the dark from falling rocks colliding.

...AND AFTER THE 14 DECEMBER 1991 ROCK AVALANCHE

Daylight revealed that the mountain had been dramatically changed. A great rock avalanche, estimated at 14 million cubic metres (18.3 million cubic yards) of the rock buttress, had travelled 7 km (4.3 miles) from the peak down the Hochstetter Icefall and Mount Cook was now 20 m (66 feet) shorter with a 500 m (1640 feet) scar. The debris from the avalanche stretched 2 km (1.2 miles) across the Tasman Glacier below the mountain. The peak remains unstable and another similar event could occur at any time.

WAKING THE GIANT

Bill McGuire

Although the slow dance of the Earth's tectonic plates across the face of the planet takes place at about the same speed that fingernails grow, the grinding encounters between these colossal slabs of rock build up immense strains within the rigid, outer carapace of our world and provide passages for hot magma from below to reach the surface. Barring Jupiter's small but staggeringly violent moon, Io, Earth is the most dynamic body in the solar system, so it should hardly be a surprise that its surface is constantly restless and often hazardous. Giant volcanic eruptions punctuate prehistory and, on one occasion around 74 000 years ago, may even have brought the human race close to extinction. Catastrophic earthquakes are described in humanity's earliest records, from accounts of the fall of Jericho to reports of battered cities in China's distant past. Tsunamis and landslides are ever present too, bringing death and destruction to coastal and mountain communities as long as humankind has been around.

Even against this background of persistent geological mayhem, however, the last decade or so stands out as a period over which the solid Earth has been unusually and dangerously violent. In particular, our world seems to have been at the heart of an earthquake storm. In less than eight years, deadly seismic events in Indonesia, Pakistan, China, Chile, Haiti, New Zealand and, most recently, Japan, have killed more than three-quarters of a million people and wrought damage totalling more than half a trillion dollars. Most extraordinarily, the events in Indonesia, Chile and Japan, were all of magnitude 8.8 or greater on the Moment Magnitude Scale, making them some of the most powerful quakes ever recorded. Such massive seismic shocks are very rare and just four of similar magnitude were recorded throughout the whole of the twentieth century. Consequently, seismologists have been racking their brains over the cause of the recent cluster. Is this a statistical fluke, natural variation, or something more? We know that a quake on one fault can stress an adjacent fault and trigger an earthquake there, but can a gigantic quake in one place promote others half a world away and years later? The jury is still out.

We have to go back more than twenty years, to 1991, for the last major volcanic eruption – at Mount Pinatubo in the Philippines. Recent eruptions have been smaller, but no less disruptive. While the Soufrière Hills eruption on Montserrat rumbles on after nearly two decades, leaving this formerly idyllic Caribbean island drenched in ash, others in Iceland and Chile have spawned air traffic chaos across the world. The 2010 eruption of Eyjafjallajökull demonstrated admirably the long reach of volcanoes, when it left the UK and Europe under a pall of ash for a week, causing the cancellation of more than 100 000 flights and disrupting the travel plans of ten million people. The following year, it was the turn of another Icelandic volcano, Grimsvötn, to again close down airways in the region. Meanwhile, in the southern hemisphere, the roaming ash clouds from the 2011 eruption of Chile's Puyehue-Cordón Caulle volcano resulted in similar travel chaos across South America, southern Africa, Australia and New Zealand.

Looking to the future, there seems little prospect of reducing the impact of the tectonic and volcanic upheavals occurring beneath our feet. The relentlessly rising global population and the concentration of more and more people in expanding megacities in geologically hazardous parts of the world will ensure that disasters keep happening and that many thousands of lives will continue to be lost. Global warming due to human activities may make matters even worse, with growing evidence that climate change in the past has elicited a violent response from the Earth and may do so again. Already, the rapid shrinking of glaciers in Alaska has reduced the weight on active faults beneath them, allowing them to move more easily, thereby increasing seismic activity in parts of the state. In mountain regions across the planet, growing numbers of heat waves are destabilizing mountain faces and promoting more massive landslides. Scientists are already speculating that the loss of Iceland's Vatnajökull Ice Cap may promote increased eruptive activity from the volcanoes beneath it, while melting of the Greenland Ice Sheet could wake up active faults with the potential to trigger earthquakes and Atlantic tsunamis. It could be, then, that a warmer world will also be a more geologically dangerous one.

Volcanic ash pours from Iceland's Eyjafjallajökull Volcano in April 2010 and drifts southwards to cause widespread disruption to air traffic across Europe. The ash spread at two levels – in a lower, diffuse plume and in a thicker layer stretching to a height of 7315 m (24 000 feet).

EXTREME

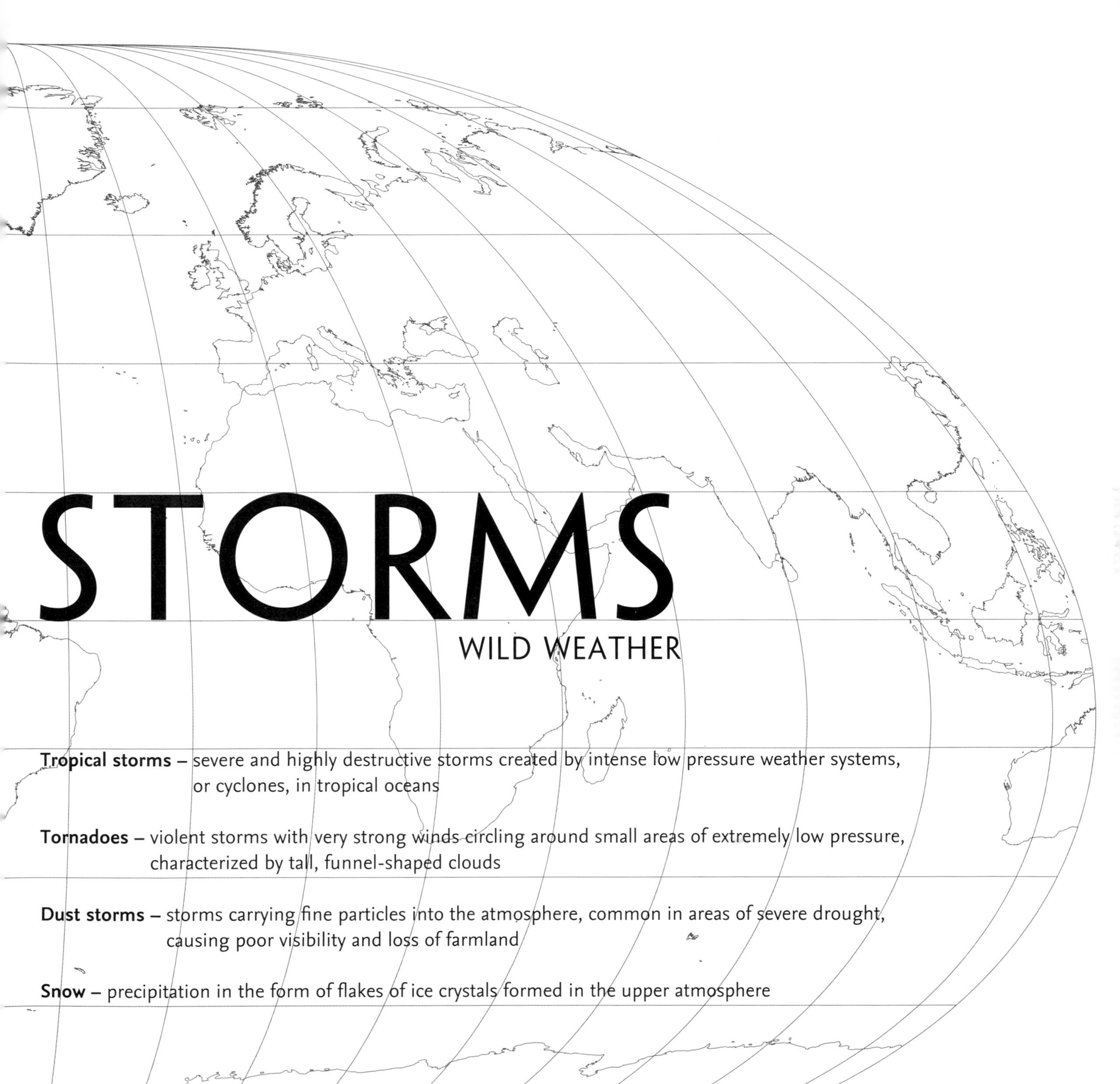

STORMS

WILD WEATHER

Tropical storms – severe and highly destructive storms created by intense low pressure weather systems, or cyclones, in tropical oceans

Tornadoes – violent storms with very strong winds circling around small areas of extremely low pressure, characterized by tall, funnel-shaped clouds

Dust storms – storms carrying fine particles into the atmosphere, common in areas of severe drought, causing poor visibility and loss of farmland

Snow – precipitation in the form of flakes of ice crystals formed in the upper atmosphere

TROPICAL STORMS –

severe and highly destructive storms created by intense low pressure weather systems, or cyclones, in tropical oceans

Caribbean Storm

TROPICAL STORMS

Tropical cyclones, known as hurricanes in the Atlantic and Caribbean, typhoons in the Pacific and China Seas, and cyclones in the Bay of Bengal, begin as a disturbance in the distribution of air pressure. If the disturbance intensifies, generating sustained winds of up to 60 km (37 miles) per hour, it is classified as a tropical depression. When the winds increase beyond that it is called a tropical storm, and given a name. It becomes a tropical cyclone when it sustains winds of more than 120 km (75 miles) per hour. As the storm crosses the ocean, water evaporates into it, condensing to produce towering clouds. Condensation releases latent heat, warming the air and making it rise further. Evaporation and condensation supply the energy to drive the storm, and to generate a tropical cyclone the sea-surface temperature must be at least 24°C (75°F). That is why tropical cyclones develop only in the tropics and only during the summer. They also need the Coriolis effect, caused by the reaction of the atmosphere to the Earth's rotation, to make them turn. The Coriolis effect does not exist at the equator. Consequently, tropical cyclones cannot form closer than five degrees to the equator. Once formed, the storm travels westward until it approaches a continent. Then its track curves away from the equator. The Coriolis effect increases with distance from the equator, turning the track further.

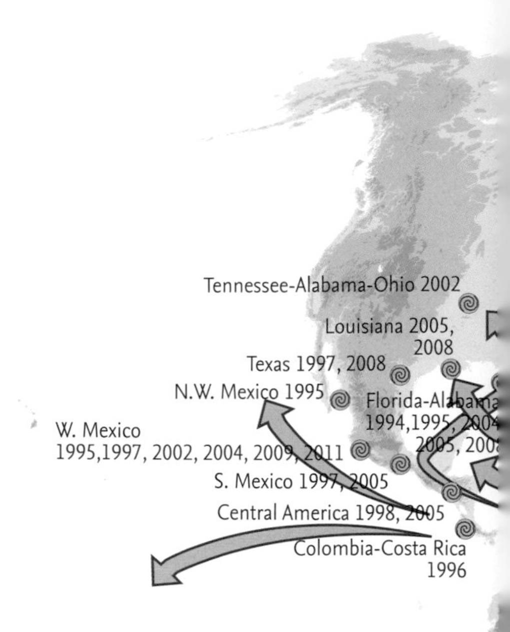

2011 Hurricane tracks

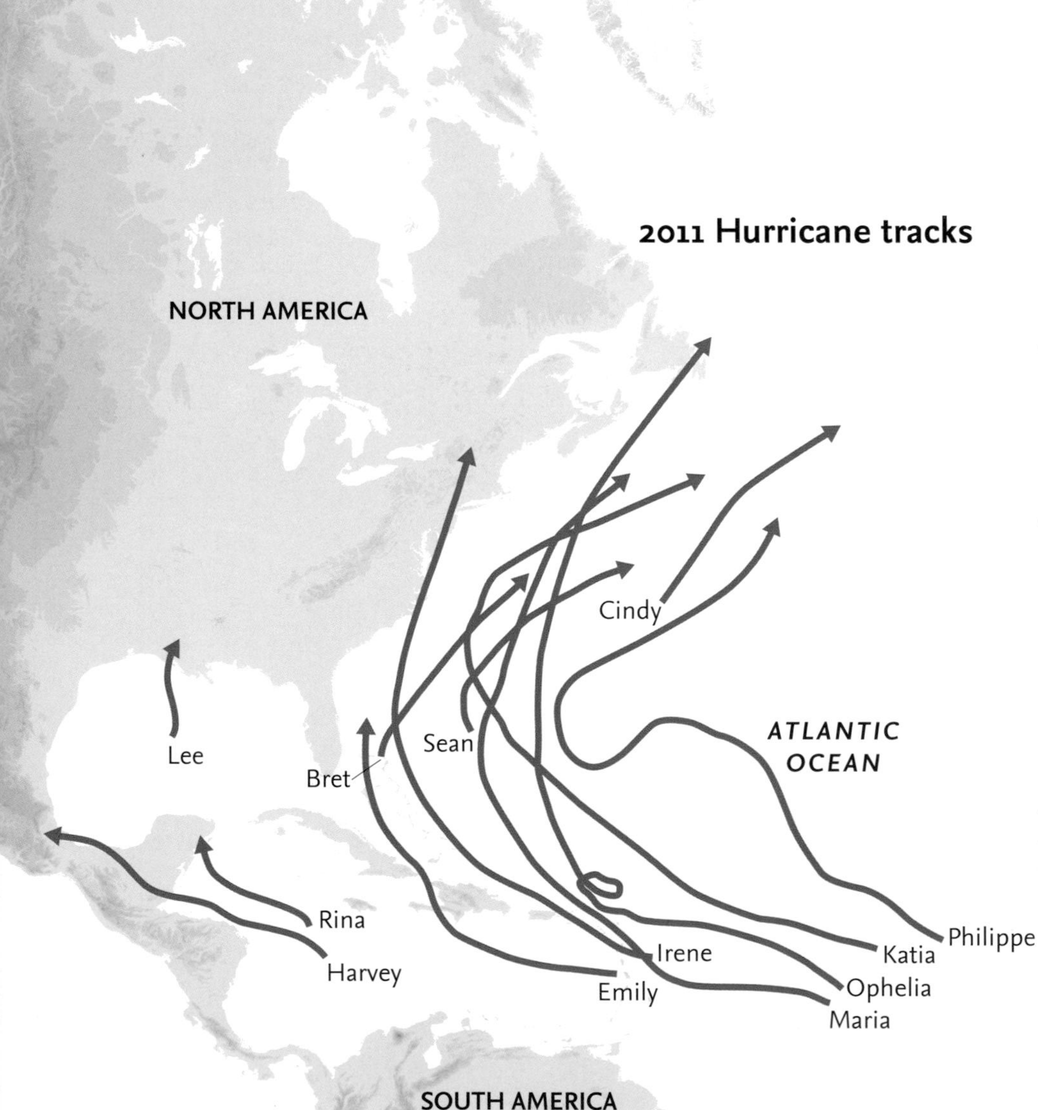

In 2011, for the fourth consecutive year, North America did not suffer a hurricane greater than force 4. Despite that, hurricane Irene caused over US $18.7 billion of damage as it swept through the Caribbean and up the northeastern states of the USA. Irene reached category 3 and touched land in nine different places causing 250-350 mm (10-14 inches) of rainfall in some areas, resulting in serious flood damage. New Orleans was wary when hurricane Lee approached but six years after the devastating damage caused by Katrina, Lee decreased in intensity to cause relatively little damage. Hurricane Katrina had caused nearly 1 400 deaths and over US $60 billion worth of damage.

Hurricanes generate ferocious winds, but it is water, not wind, which causes the greatest damage and loss of life. As well as the heavy rain, the low pressure at the core of the storm allows the sea to bulge upward and the winds drive huge waves toward the shore. Together these produce a storm surge which can cause coastal flooding. Katrina produced a 9 m (30 foot) storm surge.

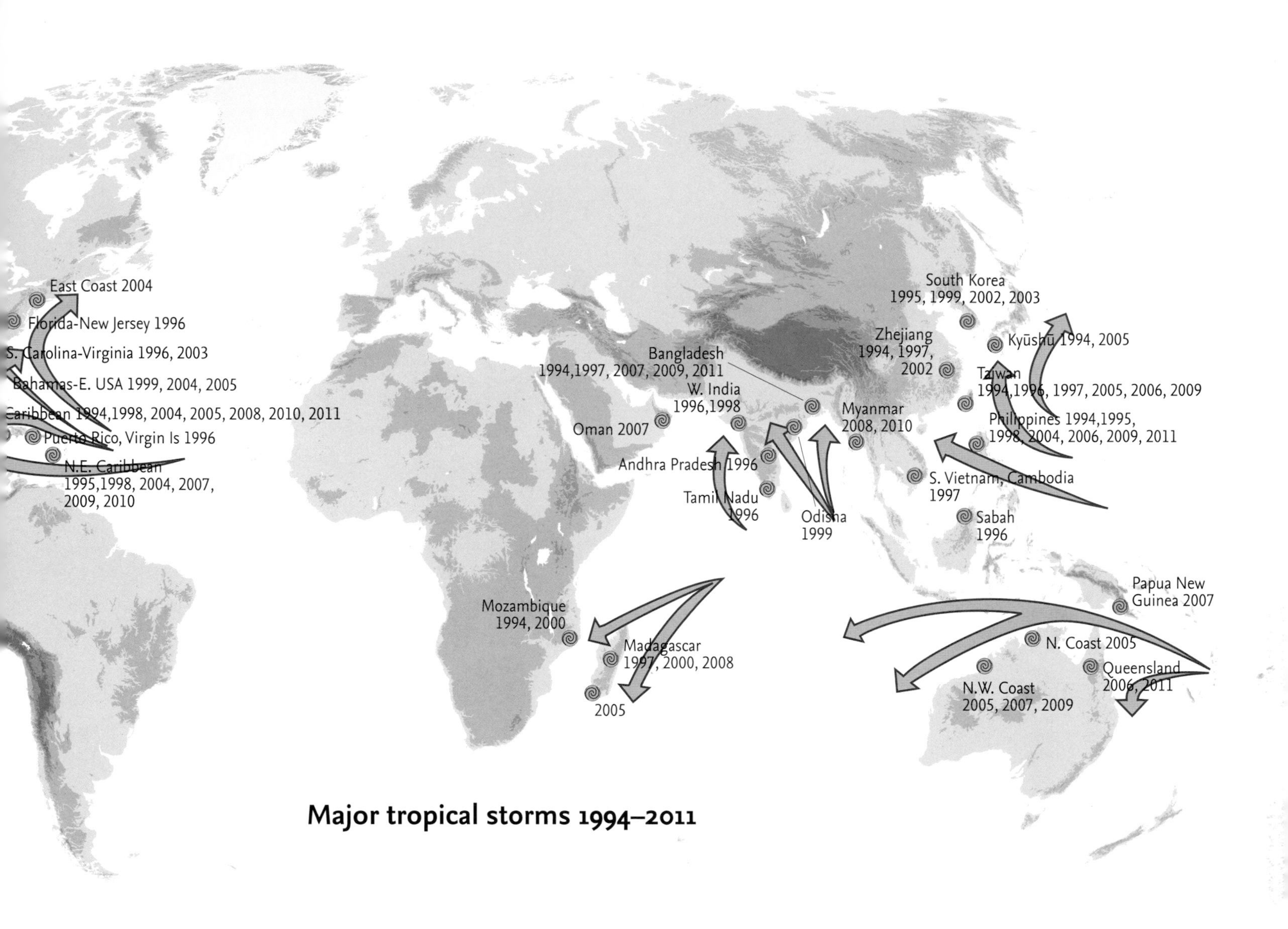

Major tropical storms 1994–2011

The Saffir-Simpson hurricane scale

Tropical storms are reliant on warm water below them for energy and the slope of the continental shelf can affect the speed they hit land. Storms will abate when passing over land. The Saffir-Simpson scale ranks hurricanes from one to five in severity.

	Pressure		Wind speed			Storm surge above normal	
Category	kilopascals	millibars (mb)	km/h	mph	knots	m	ft
Tropical storm			63–118	39–73	34–63		
1	over 98	over 980	119–153	74–95	64–82	1.2–1.5	4–5
2	96.5–98	965–980	154–177	96–110	83–95	1.8–2.5	6–8
3	94.5–96.5	945–965	178–209	111–130	96–113	2.8–3.7	9–12
4	92–94.5	920–945	210–249	131–155	114–135	4–5.5	18
5	under 92	under 920	over 249	over 155	over 135	over 5.5	over 18

Cross section of a tropical storm

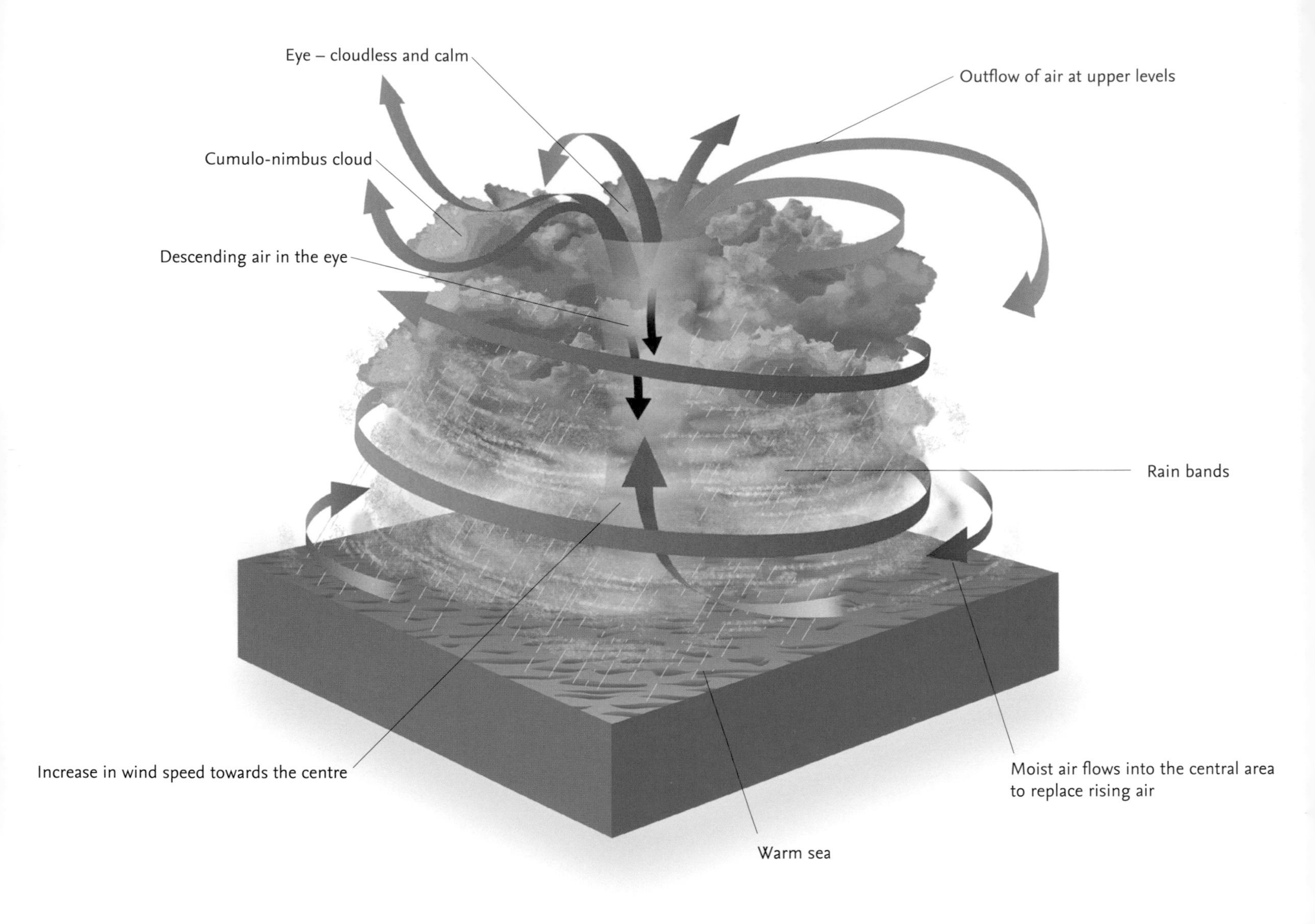

There were fifteen Atlantic hurricanes in 2005 and the official 1 June to 30 November hurricane season had to be extended into January 2006. This made 2005 a record-breaking year, with hurricane Katrina being the most destructive. But it came after a period of unusually low hurricane frequency and there have been worse. There were eighteen hurricanes in 1968 and twenty-one in 1933. The last four years saw the number of hurricanes reducing back to near the annual average. Hurricanes are probably not becoming more frequent, and the winds they produce are not growing stronger, but the proportion of hurricanes in the fiercest category is increasing. This may be due to global warming.

Storm intensity is measured as the surface atmospheric pressure at the centre. While still over the ocean, Katrina's core pressure fell to 90.2 kilopascals (902 millibars). This made it the fourth most powerful storm ever recorded in the Atlantic and Caribbean. It was not the deadliest storm to strike the United States, however. That was the one which struck Galveston, Texas, in 1900, killing between 6 000 and 12 000 people.

NEW ORLEANS, LOUISIANA, USA 28 AUGUST 2002 AND 5 OCTOBER 2005

On a fateful day in October 2005 the unthinkable happened to New Orleans. Category 3 hurricane Katrina, having abated from category 5, struck the Louisiana coast making landfall at probaby the most vulnerable point – New Orleans. The combination of high winds and torrential rain resulted in the levees being breached in a number of places, causing widespread flooding across 80 per cent of the city. The satellite images above show the inundation alongside the Inner Harbor Navigation Canal.

HURRICANE IRENE ABOUT TO MAKE LANDFALL IN NORTH CAROLINA, USA 26 AUGUST 2011

The swirl of cloud of Hurricane Irene shows the typical form of a tropical storm, with the clouds spiralling around the 'eye' in the centre. Originating in the Atlantic Ocean, Irene tracked northwest towards the Caribbean, reaching category 3 status on the Saffir-Simpson hurricane scale with winds of 185 km (115 miles) per hour. Irene weakened and swept north up the eastern coast of the United States. Irene caused fifty-six deaths, intense flooding, and left millions of people without power. It was one of the most expensive hurricanes in United States history.

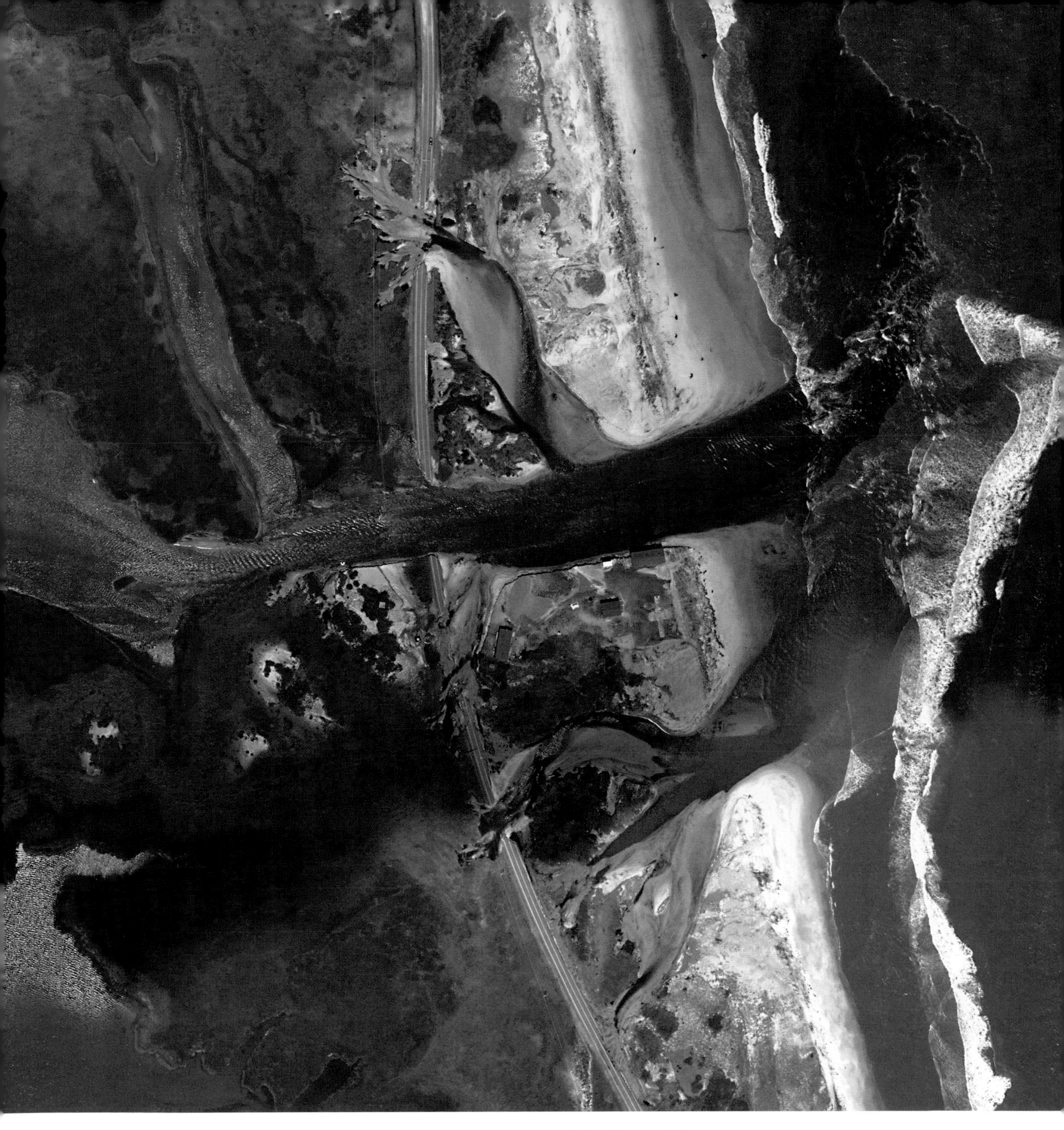

STORM DAMAGE ON HATTERAS ISLAND, NORTH CAROLINA, USA 28 AUGUST 2011

North Carolina's Hatteras Island was temporarily cut off from the mainland following intense storm damage from Hurricane Irene, stranding the barrier island's 2500 residents. Flood waters cut through numerous sections of Highway 12, which is the only road connecting the island to the mainland. The aerial photo shows several of the highway's damaged sections; water and shifting sand can be seen covering the road. It took six weeks to rebuild the road and in the meantime residents had to use ferries if they wanted to leave the island.

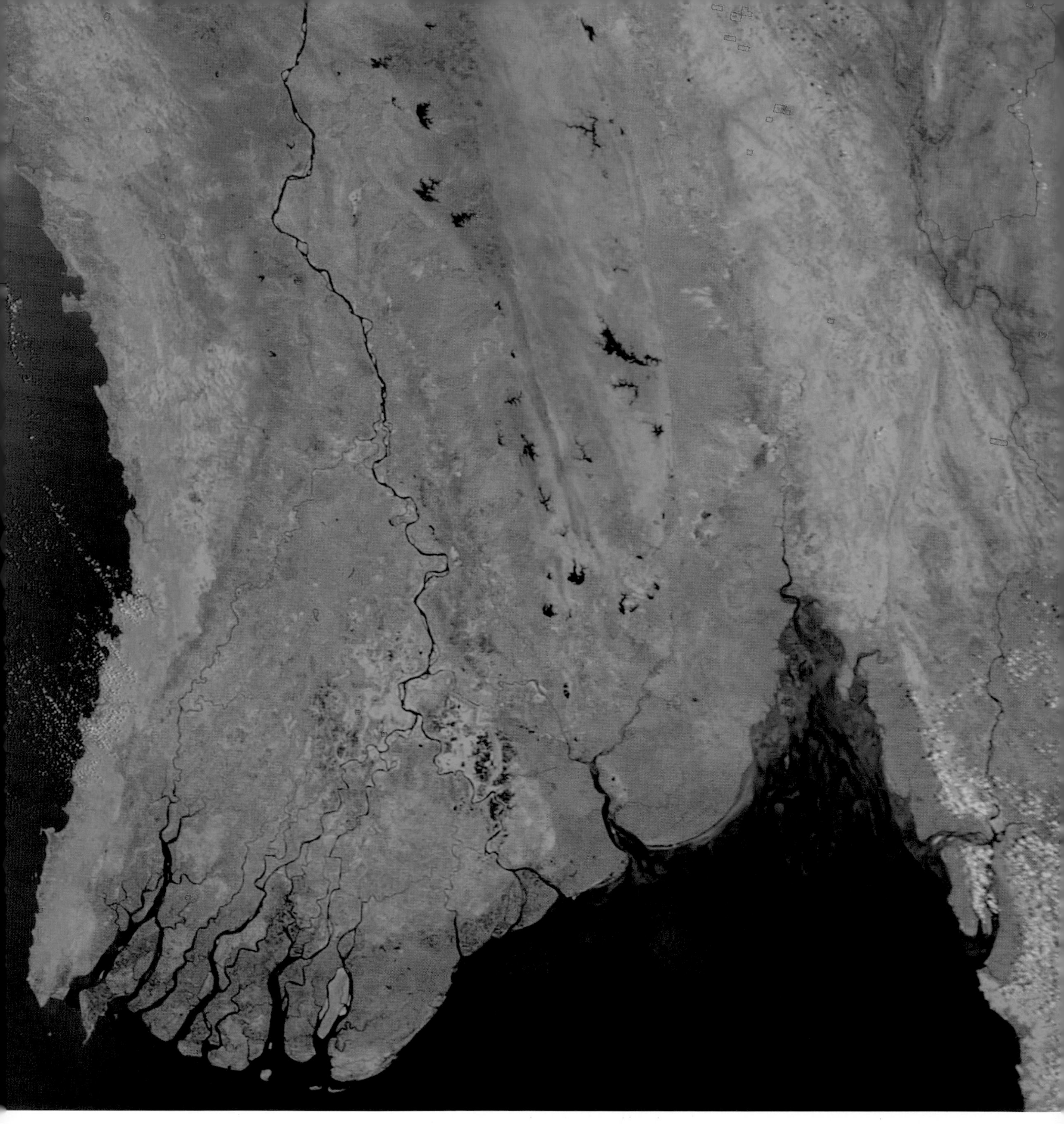

MYANMAR 15 APRIL 2008

In this satellite image lakes and rivers can be clearly seen against the background of vegetation and agricultural land. The main river, the Irrawaddy, flows south and splits into many separate channels becoming a delta known as the Mouths of the Irrawaddy. The darker blue-green colour of the coastal wetlands is also clear.

MYANMAR AFTER CYCLONE NARGIS 5 MAY 2008

Cyclone Nargis hit Myanmar on 2 May 2008 with winds of 209 km (130 miles) per hour and gusts of 241–257 km (150–160 miles) per hour which is between a Category 3 and 4 hurricane. There was widespread flooding in the country's most populated area and it is estimated that more than 138 000 people died. The extent of the flood water is quite clear in the later image with many areas, including Yangôn, badly affected.

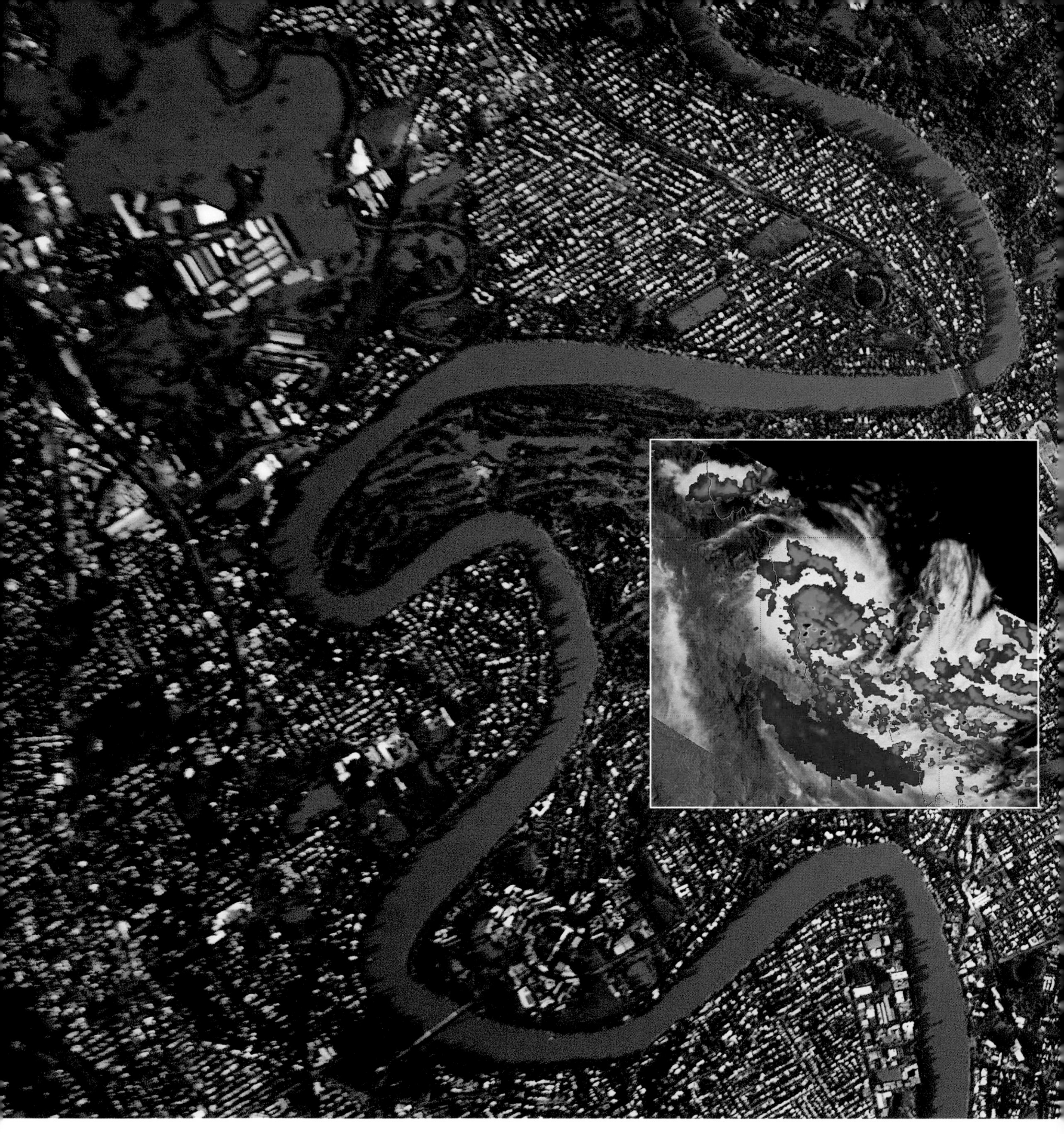

SEVERE FLOODING IN QUEENSLAND, AUSTRALIA 13 JANUARY 2011 AND 24 DECEMBER 2010

When the city of Brisbane received unusually heavy rainfall on 10 January 2011, low-lying areas suffered catastrophic flooding. This was due, in part, to the land already being heavily-saturated from Tropical Cyclone Tasha. The tropical cyclone was fairly short-lived but produced widespread torrential rain which hit Australia's northeastern coast in late December 2010. In the inset image above, precipitation data observed by NASA and JAXA's Tropical Rainfall Measuring Mission Satellite is shown. Rainfall levels show red being the highest and blue the lowest intensity.

BRISBANE RIVER, AUSTRALIA 13 JANUARY 2011 AND 9 JANUARY 2012

The series of devastating floods over the winter of 2010–2011, damaged thousands of homes and businesses across central and southern Queensland. Thousands of people were evacuated, but there were thirty-five fatalities. The flooding in Brisbane was severe, with areas around Brisbane river being the worst affected. This pair of images show a Brisbane City Ferry mooring site during the flood in January 2011 and one year later. All twenty-four of the city's ferry terminals were damaged but luckily the ferry fleet escaped harm.

TORNADOES – violent storms with very strong winds circling around small areas of extremely low pressure, characterized by tall, funnel-shaped clouds

Tornado in Parker, Colorado, USA

TORNADOES

Fujita tornado scale

F-scale number	Wind speed km/h	Wind speed mph	Type of damage done
F0	64–116	40–72	Light damage. Some damage to chimneys; branches broken off trees; shallow-rooted trees pushed over; sign boards damaged.
F1	117–180	73–112	Moderate damage. Peels surface off roofs; mobile homes pushed off foundations or overturned; moving cars blown off roads.
F2	181–253	113–157	Considerable damage. Roofs torn off frame houses; mobile homes demolished; large trees snapped or uprooted; light-object missiles generated; cars lifted off ground.
F3	254–332	158–206	Severe damage. Roofs and some walls torn off well-constructed houses; trains overturned; most trees in forest uprooted; heavy cars lifted off the ground and thrown.
F4	333–419	207–260	Devastating damage. Well-constructed houses levelled; structures with weak foundations blown away some distance; cars thrown and large missiles generated.
F5	420–512	261–318	Incredible damage. Strong frame houses levelled off foundations and swept away; automobile-sized missiles fly through the air to heights in excess of 100 m (328 ft); trees debarked; incredible phenomena occur.
F6	513–610	319–379	Winds of these speeds are unlikely

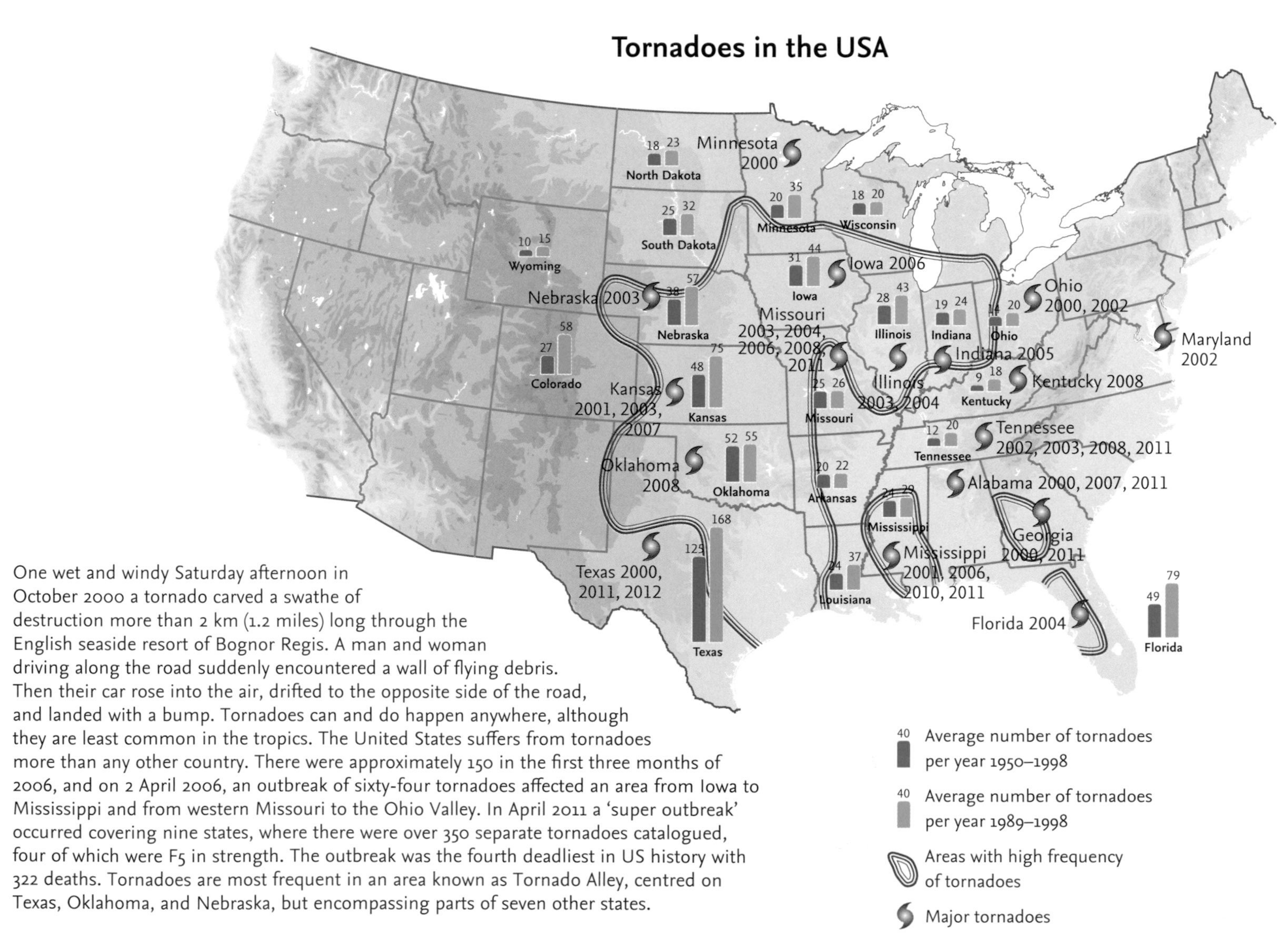

One wet and windy Saturday afternoon in October 2000 a tornado carved a swathe of destruction more than 2 km (1.2 miles) long through the English seaside resort of Bognor Regis. A man and woman driving along the road suddenly encountered a wall of flying debris. Then their car rose into the air, drifted to the opposite side of the road, and landed with a bump. Tornadoes can and do happen anywhere, although they are least common in the tropics. The United States suffers from tornadoes more than any other country. There were approximately 150 in the first three months of 2006, and on 2 April 2006, an outbreak of sixty-four tornadoes affected an area from Iowa to Mississippi and from western Missouri to the Ohio Valley. In April 2011 a 'super outbreak' occurred covering nine states, where there were over 350 separate tornadoes catalogued, four of which were F5 in strength. The outbreak was the fourth deadliest in US history with 322 deaths. Tornadoes are most frequent in an area known as Tornado Alley, centred on Texas, Oklahoma, and Nebraska, but encompassing parts of seven other states.

Major tornadoes 2000–2011

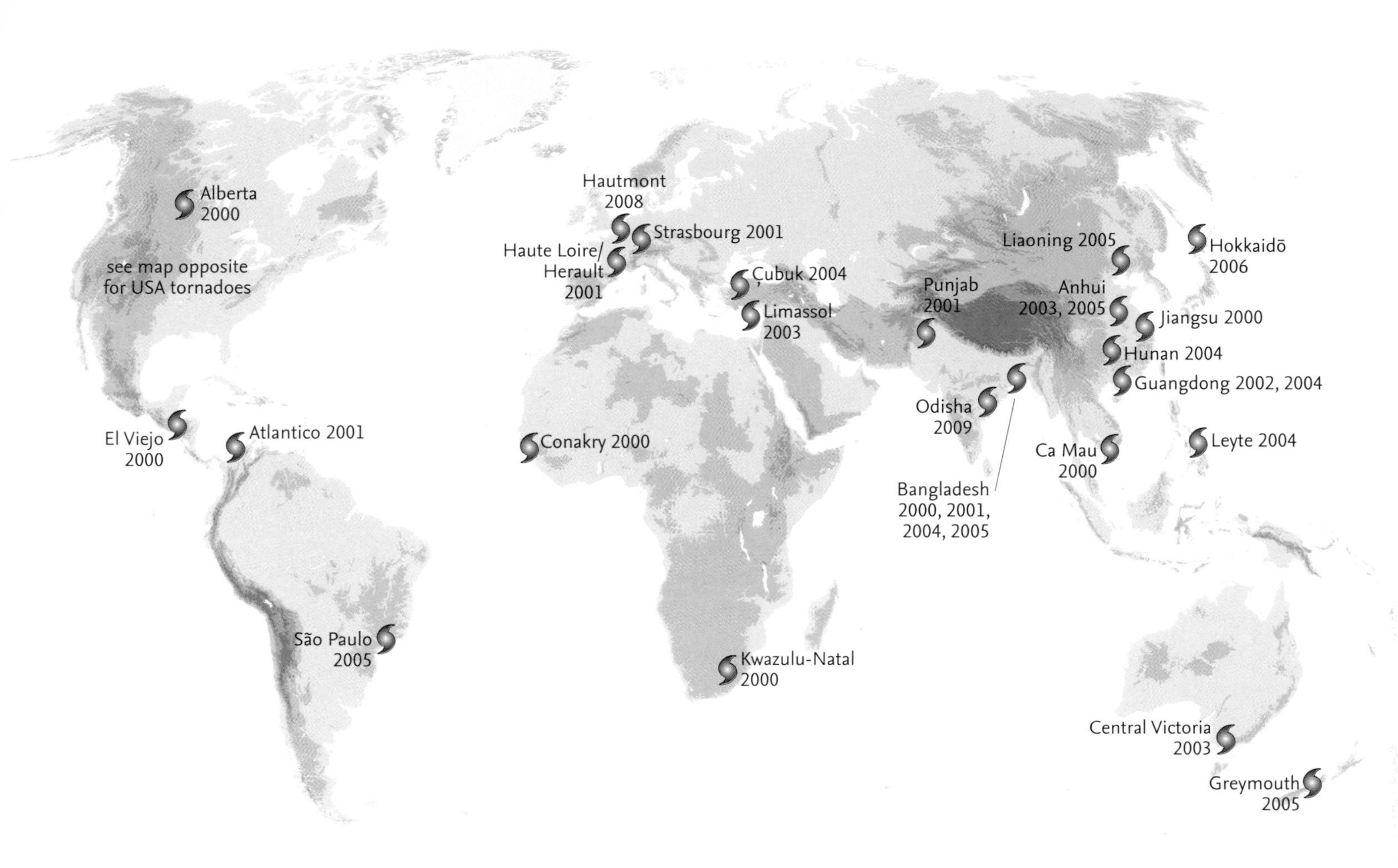

A tornado is a rapidly rotating column of air spiralling upward. The biggest and fiercest of these 'twisters' develop inside huge storm clouds containing a single convection cell, called a supercell. If air inside the cloud starts to rotate, the rotation can extend downward through the cloud, emerging from the base as a funnel, visible because the low pressure causes water vapour in the air to condense. The funnel becomes a tornado when it touches the ground and a waterspout if it touches water. Less violent conditions sometimes produce weak, short-lived tornadoes. Air accelerates as it is drawn into the tornado vortex. Most tornadoes generate winds of less than 200 km (124 miles) per hour – sufficient to cause serious damage – but occasionally the wind speed may reach 500 km (311 miles) per hour. The number of tornadoes appears to have increased dramatically over the last half century. This is due entirely to improved detection and reporting, however. There is no evidence that tornadoes are becoming more frequent.

The Fujita tornado scale was originally developed in 1971 and then updated in 1973 to take account of path length and width. It was adopted in most areas except the United Kingdom. In 2007 there was a further update and the Enhanced Fujita scale was introduced to more accurately match wind speeds to the severity of the tornado damage. The enhanced scale has not been applied retrospectively so as tornados of various eras are covered here, this book uses only the original scale for comparison purposes.

NEAR KORCULA, DALMATIAN COAST, CROATIA JULY 2004

This dramatic series of images shows the formation of a waterspout in the Adriatic Sea off the coast of Croatia. The wind speed on the edge of this waterspout reached 170 km (66 miles) per hour but it caused little damage, however, as it did not reach land. In the first image the air within the cloud has started to rotate and is emerging from it as a funnel. Gradually this funnel extends downwards until it reaches the surface of the sea and

the waterspout is formed. If this had taken place over land it would have become a tornado with a likelihood of damage occurring. In the last two images a second funnel is beginning to form. It is common for one weather system to spawn a number of tornadoes, resulting in damage over a considerable area as the system tracks across country.

TUSCALOOSA, ALABAMA, USA 27 APRIL AND 2 MAY 2011...

Between 25 and 28 April 2011, over 350 tornadoes hit the southern and eastern United States causing approximately $11 billion worth of damage and killing more than 320 people. The top satellite image shows the thunderstorms gathering over the United States. In the second satellite image, the track of the tornado which passed through the county of Tuscaloosa in Alabama on 27 April is clearly visible. Winds reached speeds of 310 km (193 miles) per hour, and the trail of desolation left by that tornado was over 129 km (80 miles) long with a maximum width of 2.4 km (1.5 miles).

… AND 29 APRIL 2011

A ground-level view of the devastation caused by the tornado which passed through the county of Tuscaloosa, Alabama, on 27 April 2011. This particular tornado caused over 1000 injuries and 65 deaths across the county. Alabama was the worst hit state out of the twenty-one that were affected by over 350 tornados which hit the United States from 25 to 28 April 2011. This was the most severe and deadly tornado outbreak to hit the United States since the 'Super Outbreak' in 1974, when 148 tornadoes hit thirteen states causing 330 fatalities and over 5000 injuries.

TORNADO DAMAGE IN JOPLIN, MISSOURI, USA 24 MAY 2011

An aerial photograph of the scene of chaos left behind by the tornado that ripped through the city of Joplin, Missouri, on 22 May 2011. The tornado rated as EF5 on the Enhanced Fujita tornado scale, and caused approximately $2.8 billion worth of damage. It was the seventh deadliest tornado in United States history, with a death toll of 161 and over 1000 reported injuries. The tornado touched down near the Kansas state line and travelled east across the city, intensifying to winds of over 322 km (200 miles) per hour as it neared the city's centre. It left a 10-km (6-mile)

long and almost 1.6-km (1-mile) wide track. Over 25 per cent of the city was completely destroyed, with the worst damage being near the centre where many buildings were totally obliterated. Across the city over 7500 residential homes and 550 businesses were totally destroyed, along with six schools and one hospital.

DUST STORMS – storms carrying fine particles into the atmosphere, common in areas of severe drought, causing poor visibility and loss of farmland

Sandstorm, Giza, Egypt

DUST STORMS

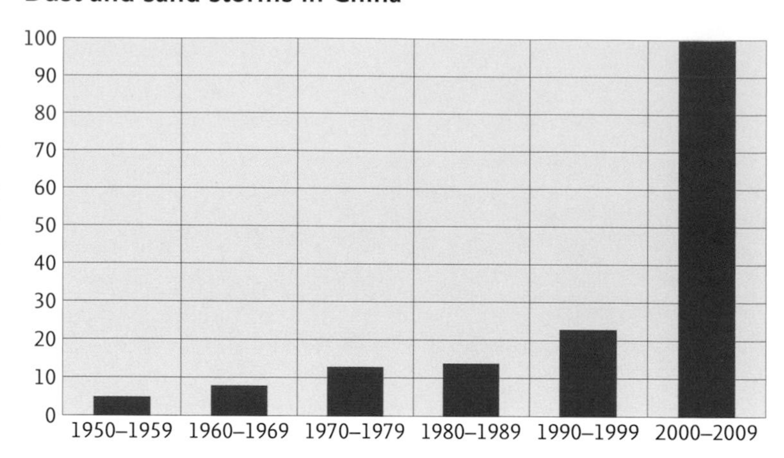

Even a light wind is enough to blow dry dust into the air and a wind of 20 km (12 miles) per hour will lift dry sand grains. Usually the dust and sand will soon settle, but if the wind blows the particles into air that is rising, they may be carried to a great height. And if the supply of dust or sand is virtually limitless, the combination of a strong wind and rising air can produce a dust storm or sand storm – the difference being only in the size of the particles. A severe storm advances as a swirling wall of dust which reduces visibility almost to zero. Dust and sand penetrate clothes and enter buildings despite the doors and windows being tight shut.

Frequent dust storm tracks

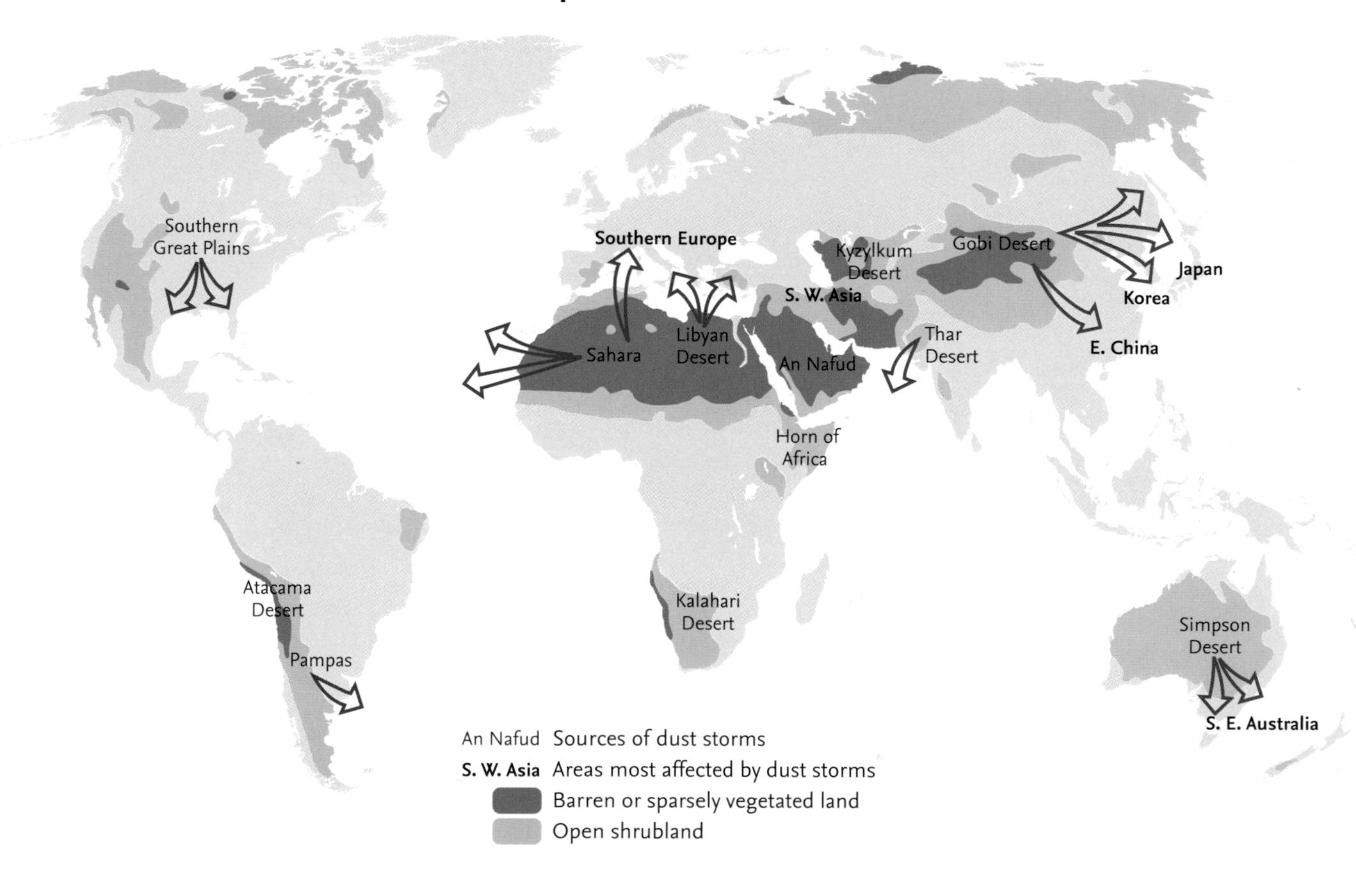

In May 1934, during the American drought which produced the Dust Bowl, dust settled on President F. D. Roosevelt's desk in Washington as fast as it could be wiped away. Ducks and geese choked and fell dead from the sky. At one time a cloud of dust – prairie soil – 5 km (3 miles) high extended from Canada to Texas and from Montana to Ohio. Once airborne, dust can travel long distances. Saharan dust quite often reaches Britain, 2 500 km (1 554 miles) away, and sometimes it crosses the Atlantic where rain washes it down onto American cars. Beijing suffers every spring when dust blows over the city from the Gobi Desert. Nearly one million tonnes of Gobi dust fall on Beijing every year. The expansion of agriculture into marginal lands and the lowering of the water table due to extraction of water for irrigation and industrial use have left the soil exposed and dry. Consequently, the Chinese dust storms are becoming more frequent and more severe. The dust spreads to neighbouring countries and as far as Hawaii. In response, the Chinese have planted millions of trees to protect the capital.

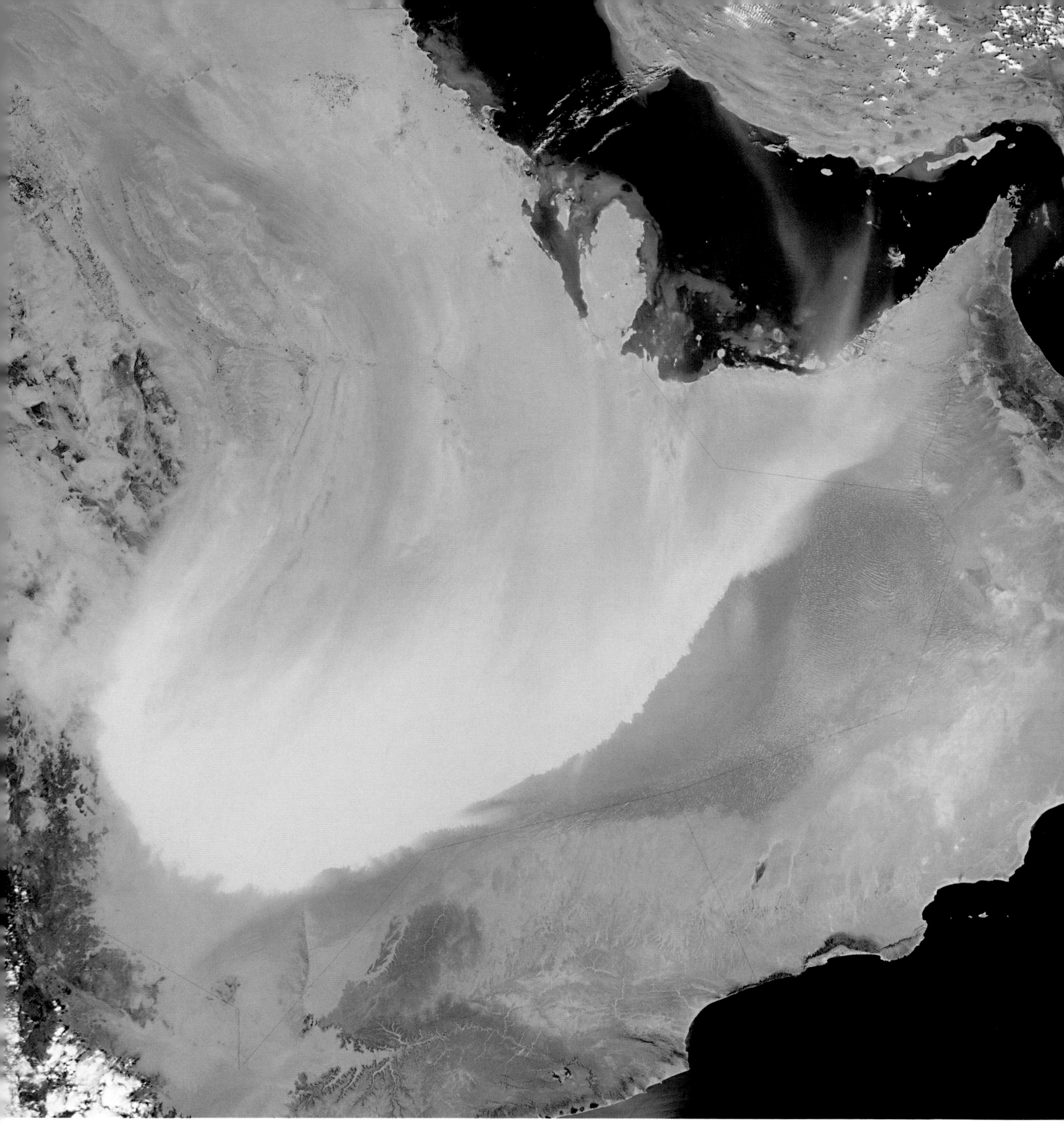

DUST STORM OVER THE ARABIAN PENINSULA 26 MARCH 2011

A massive dust storm, over 500 km (310 miles) wide, swept over the Arabian Peninsula on 26–7 March 2011. It enveloped Saudi Arabia and the United Arab Emirates, before moving south to cover most of Yemen and Oman the following day. The storm originated in Iraq and Kuwait and was driven south by northwesterly winds called 'shamal winds'. The 50 km (31 miles) per hour winds picked up sand and dust from Iraq, Kuwait and from a large sand sea called Rub' al Khālī. Visibility on the ground was reduced to less than 100 m (328 feet).

SYDNEY HARBOUR BRIDGE, NEW SOUTH WALES, AUSTRALIA

Australia is one of the world's driest continents. Dust storms are fairly common in central Australia but rarely make it to coastal areas. However, during times of extreme drought strong winds have been known to carry dust and sand all the way to eastern coastal areas. The image above is a typical view of Sydney Harbour Bridge; it differs dramatically from the image of the same bridge on the opposite page which was taken in the midst of the biggest dust storm to hit the Australian state of New South Wales in seventy years.

SYDNEY HARBOUR BRIDGE, NEW SOUTH WALES, AUSTRALIA 23 SEPTEMBER 2009

A cold front combined with strong winds whipped up dust from drought-stricken areas of central Australia before carrying it southeast to engulf the states of New South Wales and Queensland. The 100 km (62 miles) per hour winds transported 5 million tonnes of dust and sand to the coast. On 23 September the dust storm was 500 km (310 miles) wide and 1000 km (621 miles) in length. Across Sydney, transportation networks were disrupted by poor visibility; many of the city's residents suffered as air pollution levels were 1500 times their normal levels.

GARDEN CITY, KANSAS, USA 1935

Dust storms are by no means a new phenomenon. In the 1930s the Great Plains region of North America was notorious for its frequent dust storms and became known as the 'Dust Bowl'. As a result of poor farming techniques and drought the soil turned to dust and was carried away eastward by the wind, reaching as far as the Atlantic Ocean.

OBSCURED BY DUST FIFTEEN MINUTES LATER

In 1935 these two photographs were taken in Kansas, one of the many states to suffer during the Dust Bowl era. Taken only fifteen minutes apart, it is only the street lights which confirm that the images are of the same scene.

After the government promoted soil conservation programmes, the area slowly began to regenerate.

DUST STORM, PHOENIX, ARIZONA, USA 5 JULY 2011

In the summer months between May and September, dust storms are fairly common in Arizona, USA. Visibility is often reduced to near zero and air travel is disrupted until the storms pass. The image above shows a particularly impressive and major dust storm which hit the city of Phoenix in Arizona, USA on 5 July 2011. A thunderstorm to the south of the city generated high winds of over 80 km (50 miles) per hour which picked up

sand and dust from the deserts and carried the wall of dust outwards over the city itself. The wall was over 1524 m (5000 ft) high and almost 161 km (100 miles) wide by the time it reached Phoenix.

SNOW – precipitation in the form of flakes of ice crystals formed in the upper atmosphere

Avalanche in the Khumbu Icefall, Nepal

SNOW

Snow started falling earlier than usual in the UK in 2010. From 25 November heavy snowfalls continued across the country until 9 December resulting in the most widespread and significant coverage since 1965. Temperatures fell to -10°C (14°F) at night and in some places in northern Scotland, to -20°C (-4°F). After a milder spell, snow started again on 17 December, causing chaos to Christmas holiday travel. The cold north-east winds from northern Europe and Siberia caused thunder and lightning with the blizzards in some extreme cases. Figures for December showed it was the coldest month for 100 years. Heavy snowfalls stretched across into northern France, Germany and the Netherlands.

Later, in mid-February, the unusual situation occurred when all fifty states in the USA had snow in some areas, even on the tops of the Hawaiian volcanoes. Some states had not had as much snow for thirty years with as much as 300 mm (12 inches) falling in twenty-four hours.

If snowstorms are bad, blizzards are worse. Technically, a blizzard is defined as a wind of at least 56 km (35 miles) per hour with snow falling heavily enough to deposit a layer at least 25 cm (10 inches) thick or with visibility reduced to less than 400 metres (1 312 feet). People rapidly become disorientated in such whiteout conditions and may die of cold while seeking shelter. Mountain resorts welcome the snow, of course, but the quality of snow depends on the temperature inside the clouds which produce it. The best snow for skiing is dry and powdery, and falls from the coldest clouds. As global temperatures rise, however, there may be fewer winters with good snow for winter sports.

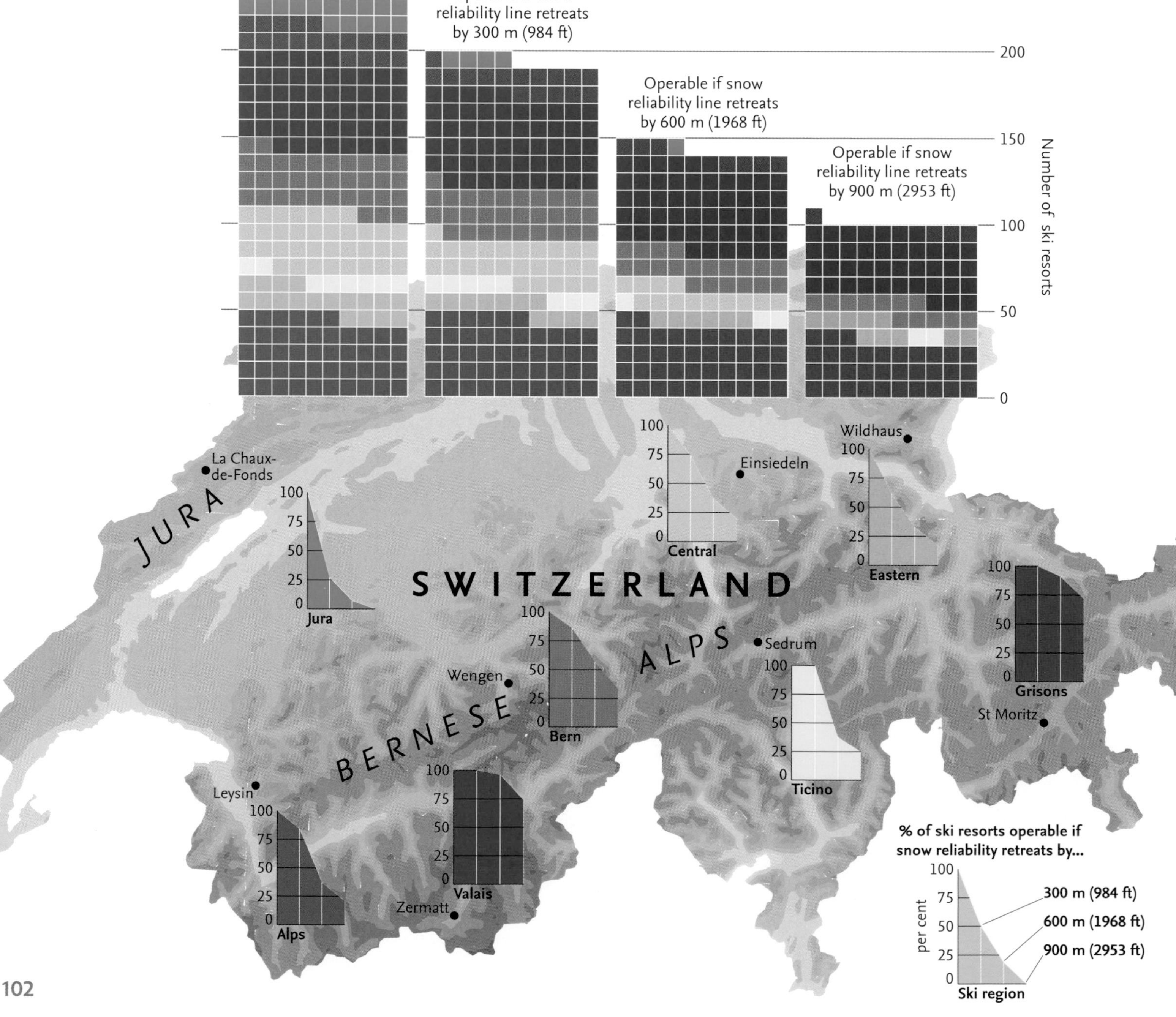

VERSOIX, LAKE GENEVA, SWITZERLAND 5 FEBRUARY 2012

In January and February 2012, Europe was plagued by sub-zero temperatures and the heaviest snowfall for decades. The cold snap was caused by a large area of high pressure that moved out of Siberia to cover most of Europe. Eastern Europe was the worst affected with temperatures dropping to -35°C (-31°F). Over 650 people died, and thousands were hospitalized due to frostbite and hypothermia. The image above shows the iced waterside promenade in Versoix. Low temperatures, combined with high winds, caused spray from the lake to cover the shore in sheets of ice.

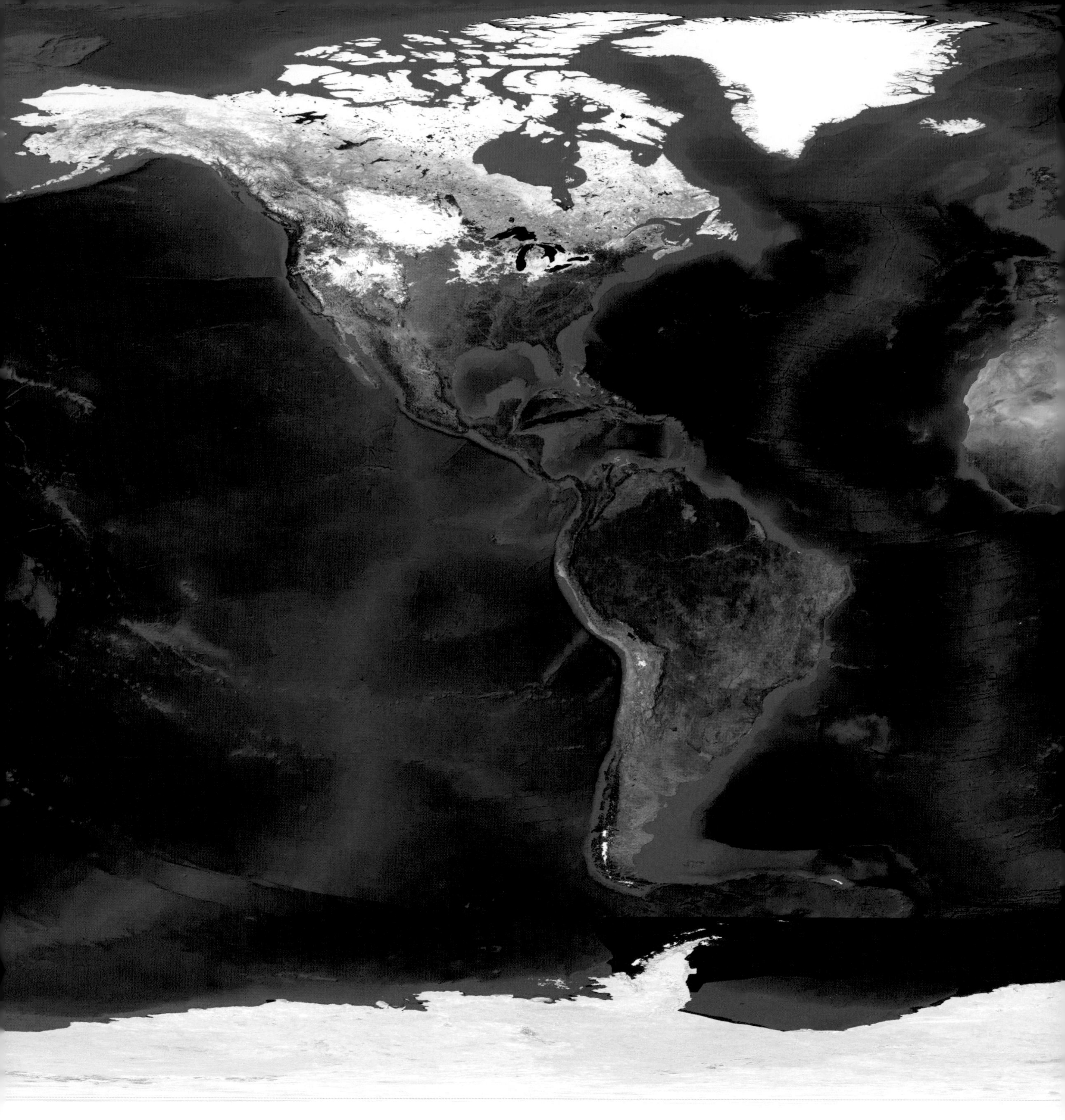

WINTER IN THE NORTHERN HEMISPHERE

Snow cover in January in the northern hemisphere extends in an almost horizontal band as far south as the USA/Canada border and from eastern Europe to northern China. South of this line the high mountain ranges of the Rocky Mountains, the Caucasus, the Himalaya and Japan are snowclad.

Western Europe, as a consequence of its position next to the Atlantic Ocean and the warming influence of the Gulf Stream, is snow-free. The southern hemisphere, apart from the Antarctic continent, is virtually clear of snow.

WINTER IN THE SOUTHERN HEMISPHERE

In July the difference in snow cover in the northern hemisphere is immediately obvious with snow now confined to the Arctic regions and the highest mountain ranges. Even though it is now winter in the southern hemisphere there is little evidence of snow except in the Andes in

South America and the Southern Alps of New Zealand. The land mass of Antarctica continues to be covered but as there is no land immmediately to its north there is no discernible extension of snow cover. The continents of Africa and Australia are almost entirely snow-free throughout the year.

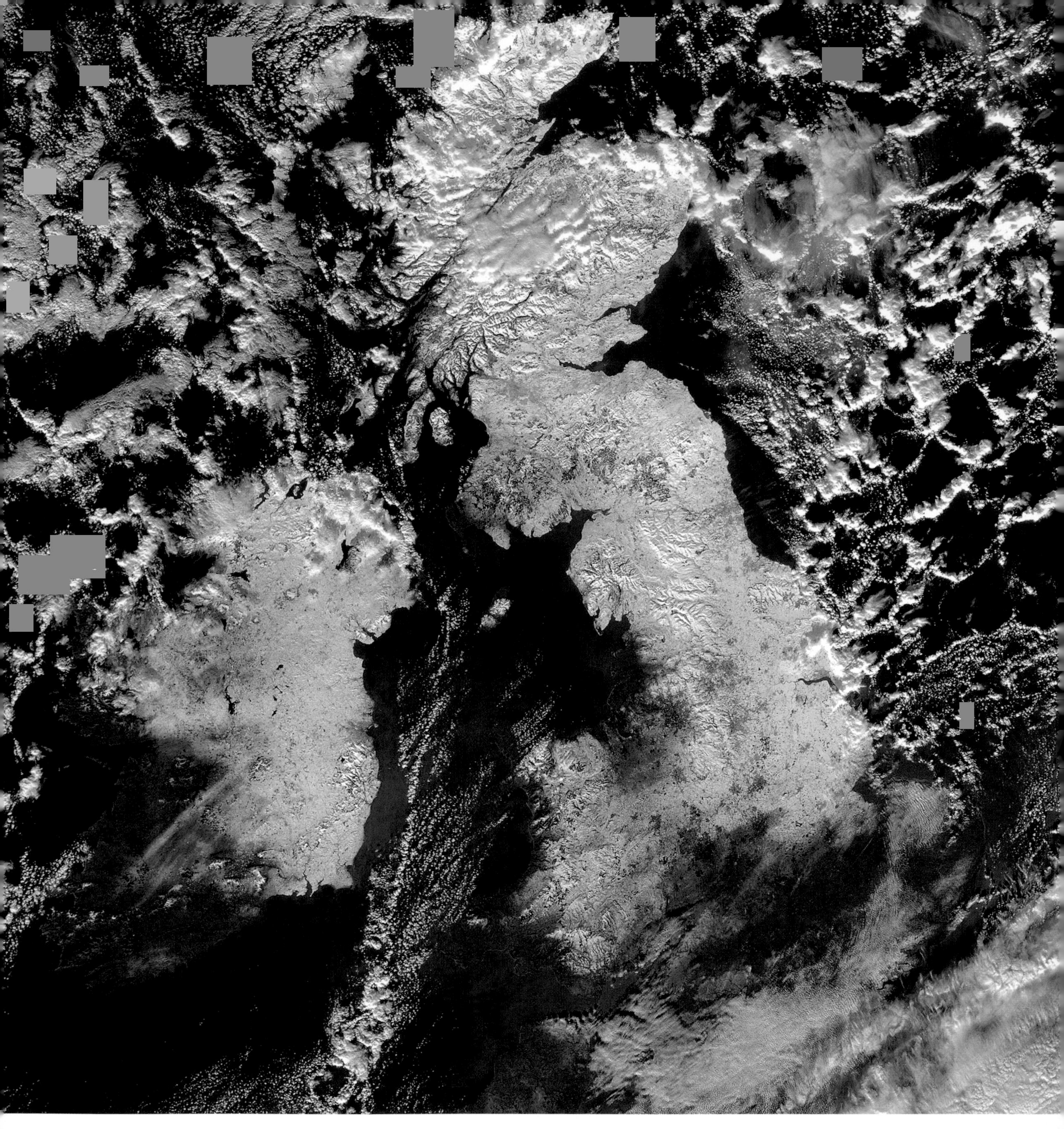

SNOW COVERS THE UNITED KINGDOM 8 DECEMBER 2010

In November and December 2010, northeasterly winds from northern Europe and Siberia caused the UK to experience its worst winter since 1962-3. Constant heavy snowfall coupled with severe night-time temperatures, which dipped to -20°C (-4°F) in Scotland, meant that temperatures struggled to rise above freezing during the day. In some areas, snow accumulated to depths exceeding 50 cm (20 inches). The satellite image shows the snow cover across the UK on 8 December, during a short-lived thaw period; a second cold spell started on 16 December.

TRAVEL CHAOS ACROSS THE UK 18 AND 19 DECEMBER 2010

The heavy snow caused chaos: schools, work places and airports were closed for days at a time, and travel by road and rail was severely disrupted. On 6 December, particularly heavy snow across the Central Belt of Scotland left hundreds of motorists stranded overnight in their vehicles on the M8, M74 and A9. The two photos above show ground-level views of the disruption caused by the snow; the first from 18 December, shows gridlocked traffic on the A3, which runs from Portsmouth to London; whilst the second from 19 December, shows several grounded planes at Gatwick airport.

ITALY 13 FEBRUARY 2012

Italy did not escape from the polar conditions that hit most of Europe throughout January and February 2012. The satellite image was captured by the European Space Agency's Envisat satellite using its MERIS instrument on 13 February 2012. It shows northern and central areas of Italy blanketed in snow in what was the heaviest snowfall event that the country had seen in decades. Heavy snow is common in the winter in northern and alpine areas of the country, but it is very rare in more southern regions such as Tuscany, Lazio and Campania.

SNOW IN ROME, ITALY 4 FEBRUARY 2012

The images above show two of Rome's most famous historic landmarks, the Colosseum and the Roman Forum, covered in snow following the heaviest snowfall event to occur in Rome in twenty-six years. Snow reached a depth of 20 cm (8 inches) and brought the capital to a standstill; there was widespread travel disruption on the roads, railways and at airports, and there were also powercuts across the city. The snow also damaged some structures, including the iconic Colosseum; several fragments of the Roman amphitheatre fell, and the structure was temporarily closed to the public.

FUTURE STORM

Bill McGuire

Amongst his many other contributions to the English language, the great American writer, Mark Twain, dreamt up definitions for weather and climate that have yet to be bettered for succinctness and accuracy: 'climate is what we expect; weather is what we get'. To put it another way, climate is the bundle of meteorological conditions – including wind, precipitation and temperature – characteristic of a particular location, while weather is what we see when we look out of the window. Where extreme storms are concerned the two are inextricably linked, particularly now, when anthropogenic climate change (changes influenced directly by man's activities) is starting to have an impact on day-to-day weather.

As the Intergovernmental Panel on Climate Change (IPCC) highlights in its recent report on extreme events, the weather is changing as the planet heats up. Since 1950, the trend has, hardly surprisingly, been towards more warmer days and nights and fewer cold ones, while heat waves have become more common and are lasting longer. Torrential rains are more frequent in some places than half a century ago, while other areas experience drought conditions more often than they used to. On the subject of the wild weather wrought by extreme storms, however, the picture is blurred. It does seem as if the tracks of powerful wind storms outside the tropics are slowly edging polewards, but on changes in the characteristics of tornadoes and tropical storms, the IPCC report is ambivalent.

Recent years have seen the US heartland taking serious beatings from swarms of tornadoes. In April 2011, the largest ever tornado swarm took more than 300 lives across six states – the deadliest episode for over three-quarters of a century. As yet, however, this does not seem to be part of a rising trend in US tornado activity. In fact, records suggest that there has not really been much change in the annual number of very powerful tornadoes since 1950.

Dwarfing any damage arising from tornadoes, the battering that New Orleans suffered in 2005 at the hands of Hurricane Katrina was a wake-up call to the US Government, the country's emergency managers and its people – particularly those inhabiting the eastern and Gulf coasts where hurricanes commonly make landfall every year. Far from being the biggest storm to strike the US, Katrina still shattered the record books with losses of more than $80 billion making it by far the most expensive natural catastrophe in US history. Inevitably, the resultant flooding had barely started to subside before connections were being made with climate change. Was this a disaster manufactured, ultimately, by the hand of man? The short answer is that we don't know. It is always going to be difficult, if not impossible, to finger a particular event and prove beyond doubt that it was the result of anthropogenic climate change. However, some statistical manipulations have made good arguments for recognizing a part played by climate change in the British floods of 2000 and the 2003 European heat wave.

Where tropical storms are concerned, though, making such a link is particularly difficult. This is because their numbers follow natural cycles controlled by annual and decadal weather and climate conditions in the Atlantic and Pacific. This complicates any trend arising from global warming. To make the situation more complicated, the incompleteness of early records makes it difficult to say whether or not numbers are increasing over time or not. Looking at the global picture there is no convincing evidence of a rise in the numbers of tropical storms. Storms do, however, seem to be generating more rainfall, thereby increasing their capacity to cause major floods, and, most importantly, there seem to be more of the really powerful storms with the potential to cause major destruction and serious loss of life.

As our world continues to warm, the trend towards more extreme storms becoming more common, even as total numbers may actually fall, brings the prospect of future catastrophe, particularly should such storms score direct hits on vulnerable megacities in the developing countries of South and South East Asia, including Mumbai and Kolkata in India and Manila in the Philippines. In 1979, the most powerful tropical cyclone ever – Typhoon Tip – rampaged across the Pacific Ocean. At its height, the storm was more than 2 000 km (1 243 miles) across, with peak wind speeds of 305 km per hr (close to 200 mph), but fortunately it had lost much of its power by the time it made landfall on the Japanese coast. A future world in which such storms were more common would be a very scary world indeed.

August 2006 saw three major tropical storms developing in quick succession in the western Pacific Ocean. Typhoons Saomai (lower right), Maria (upper right) and Bopha (left) formed over a period of four days, threatening the coasts of China and Japan. The image illustrates well the stages of development of a tropical storm.

MAN-MADE

WORLD

BUILDING AND DEVELOPMENT

Creating land – the process of reclaiming land from the sea, or converting the use of derelict or flooded areas, to create productive land

Controlling water – use of water resources for the generation of power and for the irrigation of otherwise infertile land for agriculture

Expanding cities – outward and upward growth of urban areas largely in response to the migration of people from rural areas

CREATING LAND –

the process of reclaiming land from the sea, or converting the use of derelict or flooded areas, to create productive land

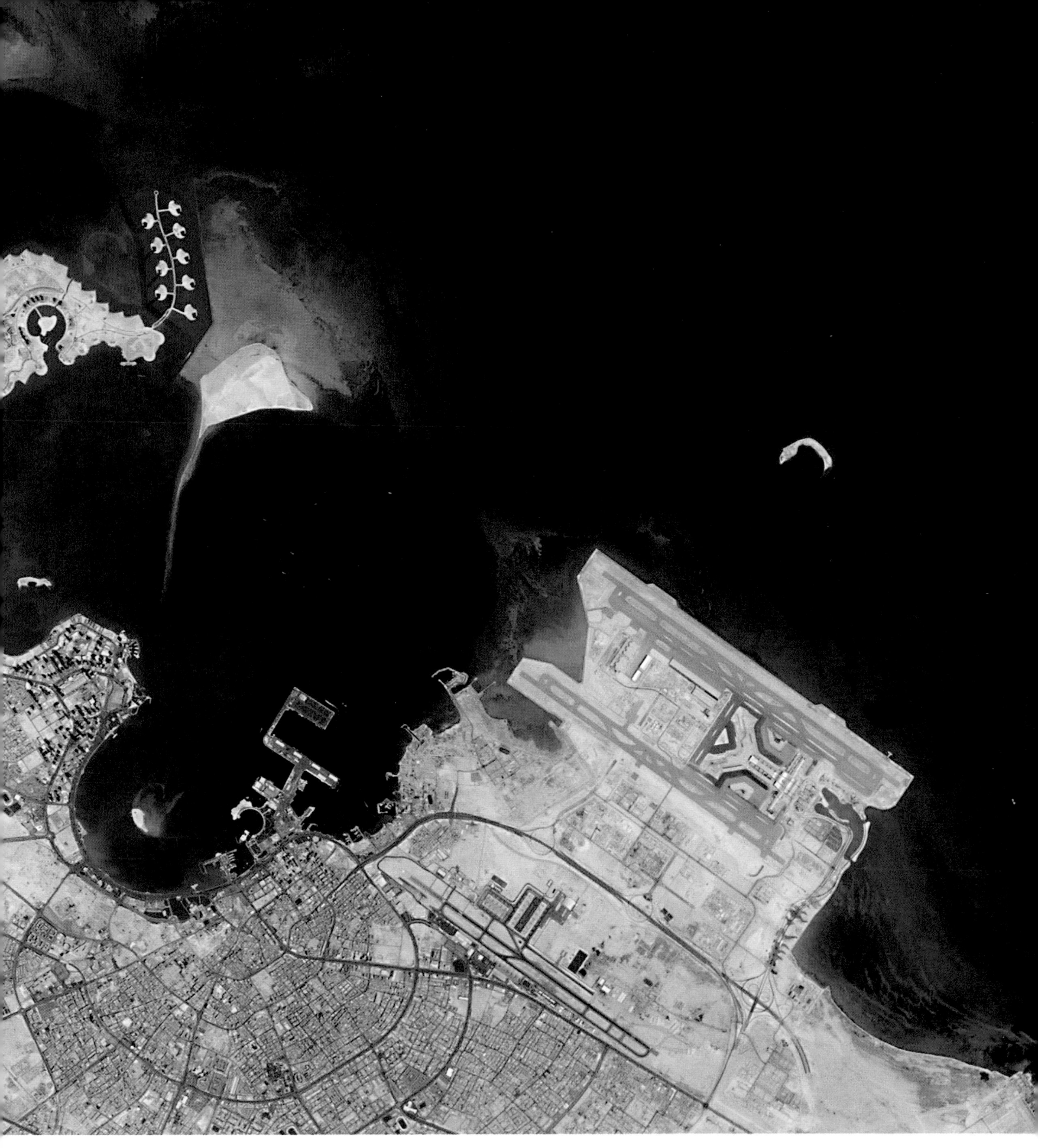

Doha and The Pearl, Qatar

CREATING LAND

Netherlands Zuider Zee project

Ijsselmeer polders	Area sq km	Area sq miles	% of total land area of Netherlands	Period of construction	% land use Agriculture	Woodland	Residential	Other
Wieringermeer	200	48	0.6	1927–1930	87	3	1	9
Northeast Polder	480	115	1.4	1937–1942	87	5	1	7
Eastern Flevoland	540	130	1.6	1950–1957	75	11	8	6
Southern Flevoland	430	103	1.3	1959–1968	50	25	18	7
Total	1 650	396	5.0	1927–1968	73	12	8	7

Two-thirds of the planet is covered by sea. Every year sea-level rise takes a little more. But in some places mankind is fighting back by draining lagoons and filling in shallow coastal waters with dredgings and waste materials. The first modern masters of land reclamation were the Dutch. A quarter of the Netherlands' land area is below sea level, mostly made up of 3 000 reclaimed 'polders'. The largest, including Flevoland, the world's largest man-made island, are within the IJsselmeer, a large lake created in the 1930s by barricading off a large bay, the Zuider Zee, from the North Sea.

Cities built on drained coastal lagoons, swamps and harbours include Venice, St Petersburg and parts of Boston and Tōkyō. Most recent large reclamation projects have been in Asia. Hong Kong has created new flat land in its harbour which now houses much of the central business district. It also recently extended Lantau Island to accommodate its new international airport. Nearby Macao airport is also built on former sea bed, as is Japan's second-largest international airport at Kansai near Ōsaka.

China has constructed the world's largest container port on reclaimed land round an island off Shanghai. The island state of Singapore has reclaimed land around sixty offshore islands, creating over 100 sq km (39 sq miles) of new land, though at the loss of several valuable coral reefs. Most dramatically, the fast-growing city of Dubai is currently building entire archipelagos of small artificial islands in The Gulf. Made from material dredged from the Gulf shipping lanes, these archipelagos will house luxury hotels, beach-front homes and holiday villas, and marinas.

But not all land reclamation goes smoothly. Kansai airport in Ōsaka, Japan is sinking. Malaysia has appealed to the United Nations over Singapore's 'reclaiming' of parts of the narrow Johor Strait between them. And the Netherlands is contemplating giving some of its polders back to the sea, because of the cost of keeping them drained. The cost both financially and ecologically to reclaim land is substantial. To reclaim one square kilometre of land takes 37.5 million cubic metres (1 324 million cubic feet) of sand.

Singapore - growth

	Area	
Date	**sq km**	**sq miles**
1963	580	224
1988	625	241
2003	697	269
2010	710	274
Future target (2030)	780	301

Singapore island land reclamation

Image of Singapore in 2004. 1958 coastline superimposed in pink.
Alterations to coastline as of 2009 in red.

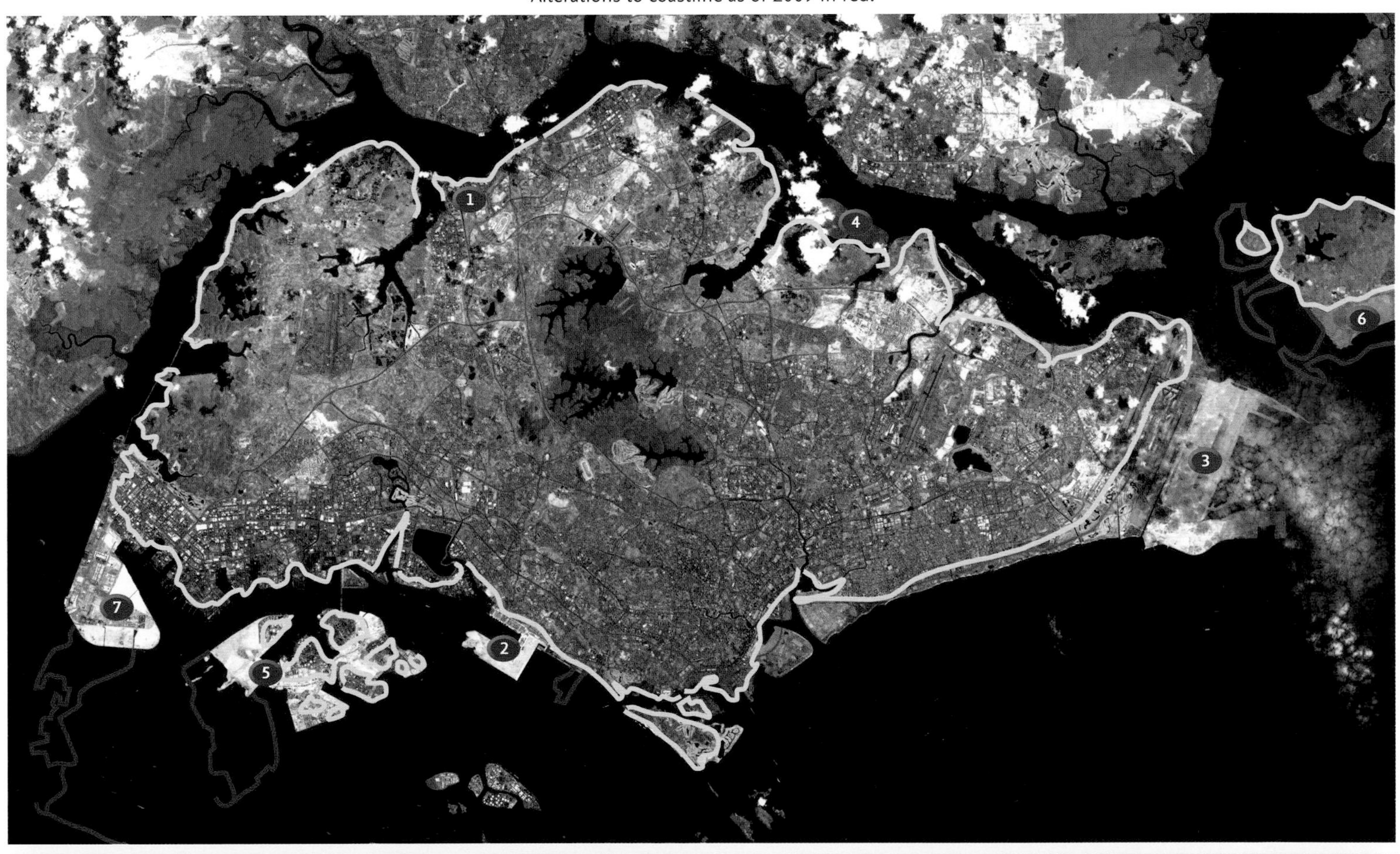

Map location	Decade	Area	Principal land use
1	1960	Kranji	industrial
2	1970	Pasir Panjang	port
3	1970	Changi	international airport
4	1980	Seletar	housing
5	1990	Jurong Island	petrochemicals
6	2000	Pulau Tekong	housing
7	2000	Tuas	industrial/port

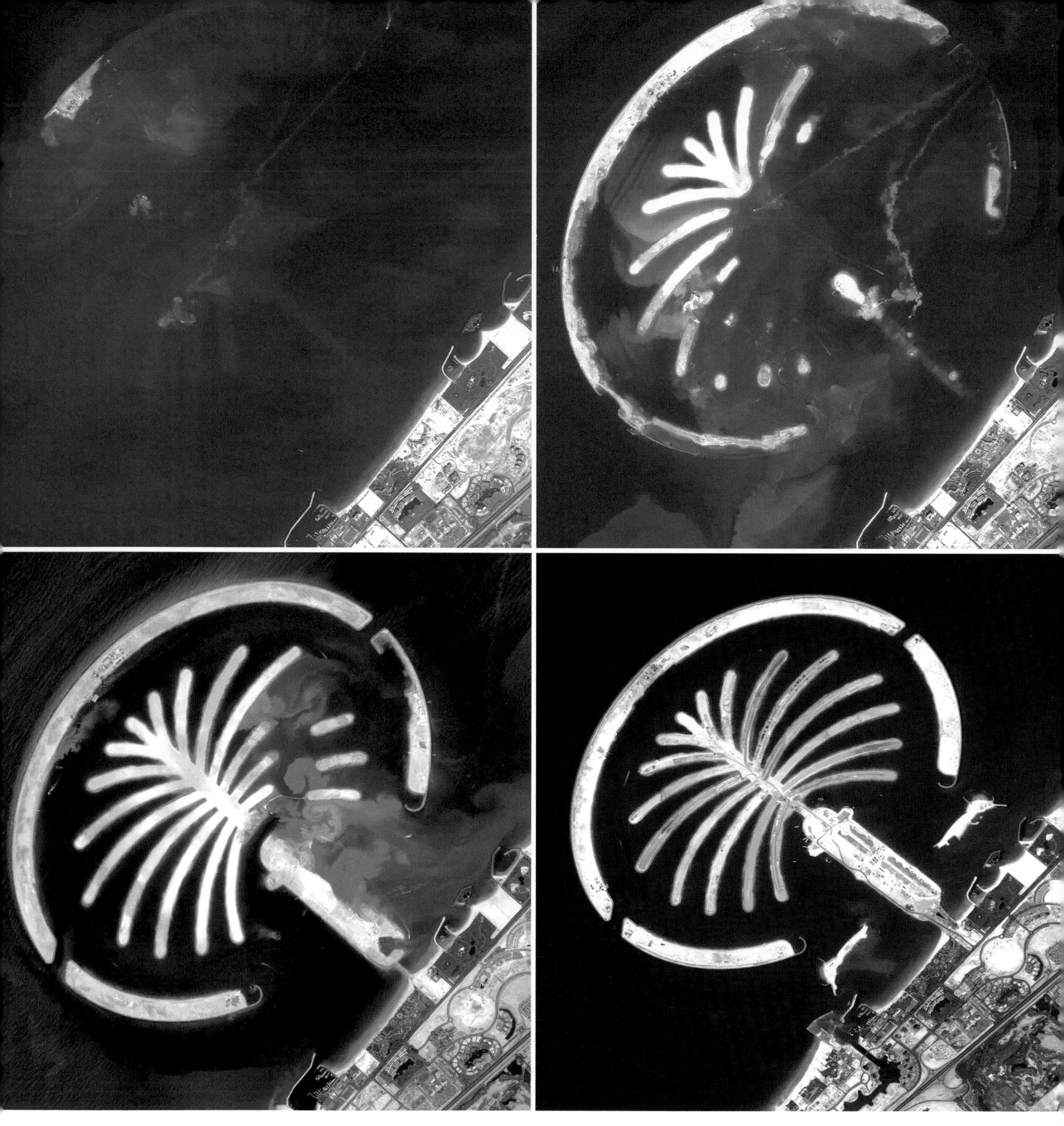

PALM ISLAND, DUBAI FEBRUARY 2002, NOVEMBER 2002, JUNE 2003, SEPTEMBER 2004

Work commenced at the beginning of the millennium on the world's largest land reclamation project to create a series of man-made islands which would become luxury resorts on the coast of Dubai in the United Arab Emirates. This sequence of images illustrates how the first 'Palm Island', Palm Jumeirah, gradually took shape as reclamation progressed. When complete, the 'island' will contain 2000 villas, 45 luxury hotels, shopping complexes and other facilities.

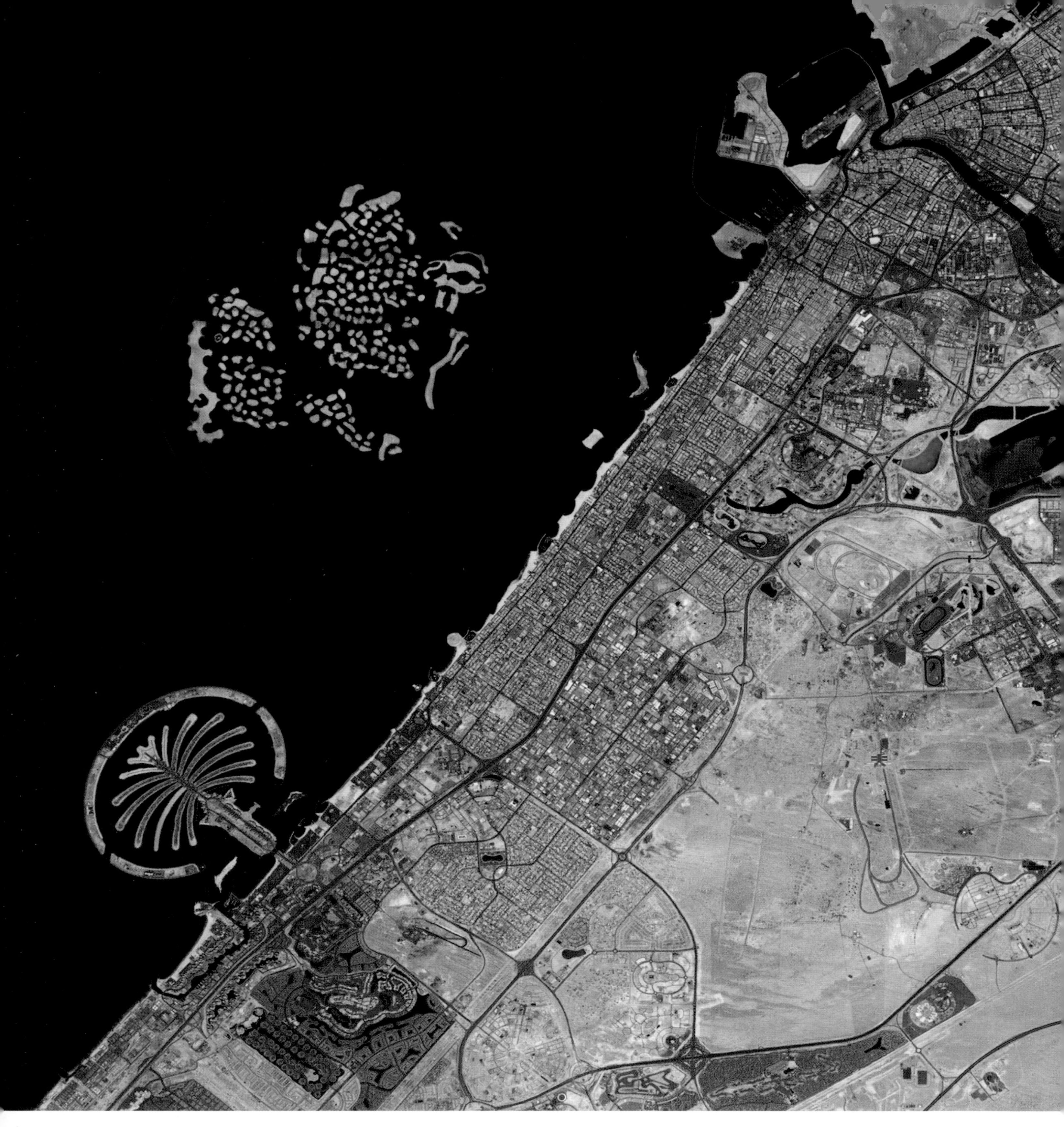

PALM ISLANDS AND 'THE WORLD', DUBAI APRIL 2011

But that was only the first part. This image taken in 2011 shows Palm Jumeirah almost complete, and the start of the third and largest of the palms, Palm Deira, at the top of the picture. However, even these developments seem small when compared to 'The World', a series of 300 islands which together give the appearance of the continents. At present only one island has been been built on. Transportation between individual islands is by sea and air.

WEST KOWLOON, HONG KONG 1997 AND 2012

Hong Kong has been reclaiming land from the sea for over 50 years to accommodate an ever-increasing population and expanding commercial activity. One of the most recent developments has been in West Kowloon where the land has been reclaimed for a variety of uses including commercial, residential and recreational. In the lower image the road link to the new international airport, built on reclaimed land off neighbouring Lantau Island, can be seen entering the tunnel under Victoria Harbour. The International Commerce Centre building now dominates.

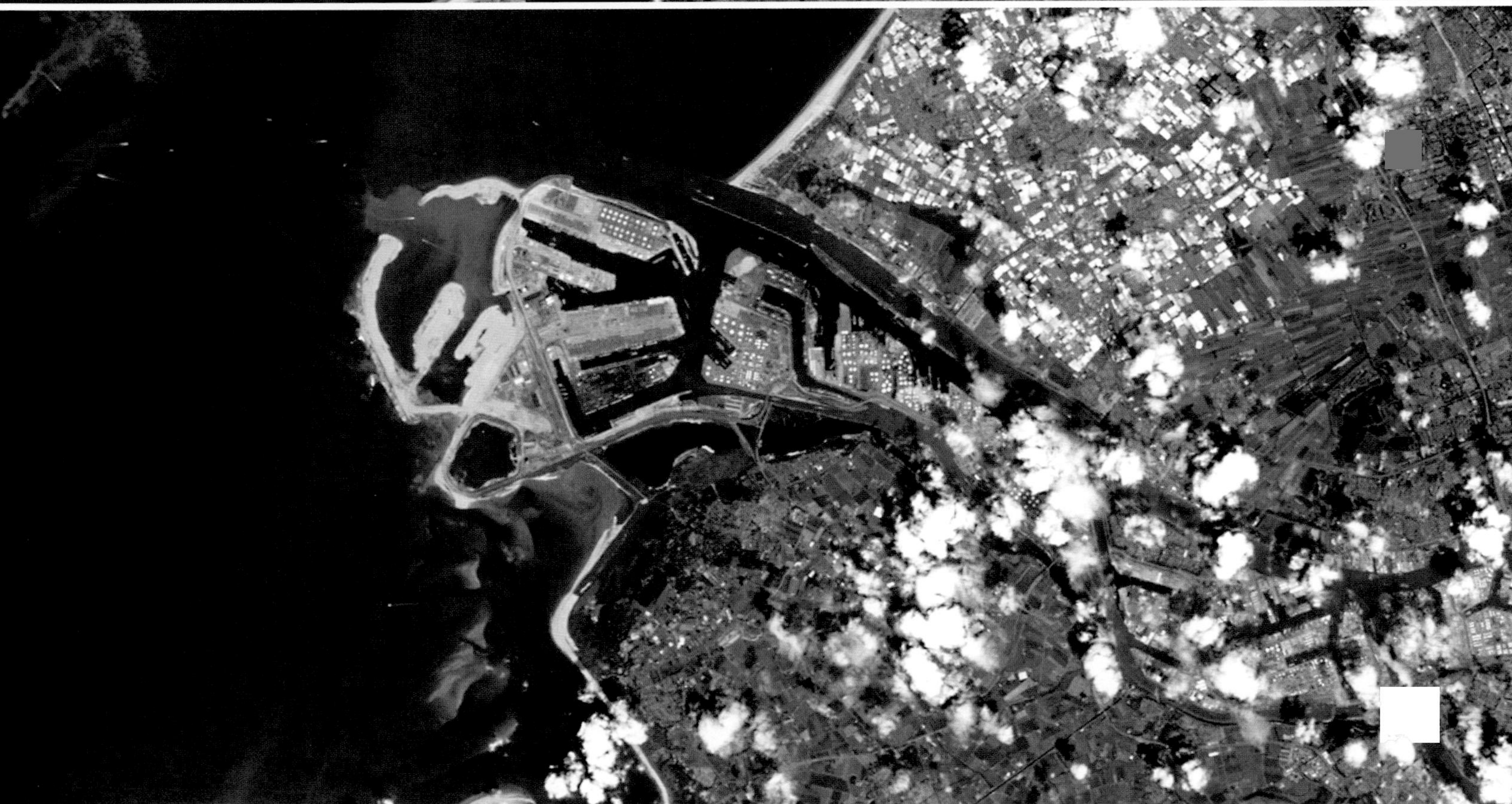

LAND RECLAMATION AT THE PORT OF ROTTERDAM, NETHERLANDS 16 JULY 2006 AND 4 JULY 2010

By 2013 Rotterdam, Europe's largest port and the third largest in the world after those in Shanghai and Singapore, will have reached its capacity. The port is vital to the Dutch economy and, in order to remain competitive, a land-reclamation project to expand the existing port by 20 sq km (8 sq miles) was started in 2008. Maasvlakte 2, the largest land-reclamation project in the Netherlands since the 1960s, will triple the port's capacity. The pair of satellite images show the port before the project began and its progress four years later.

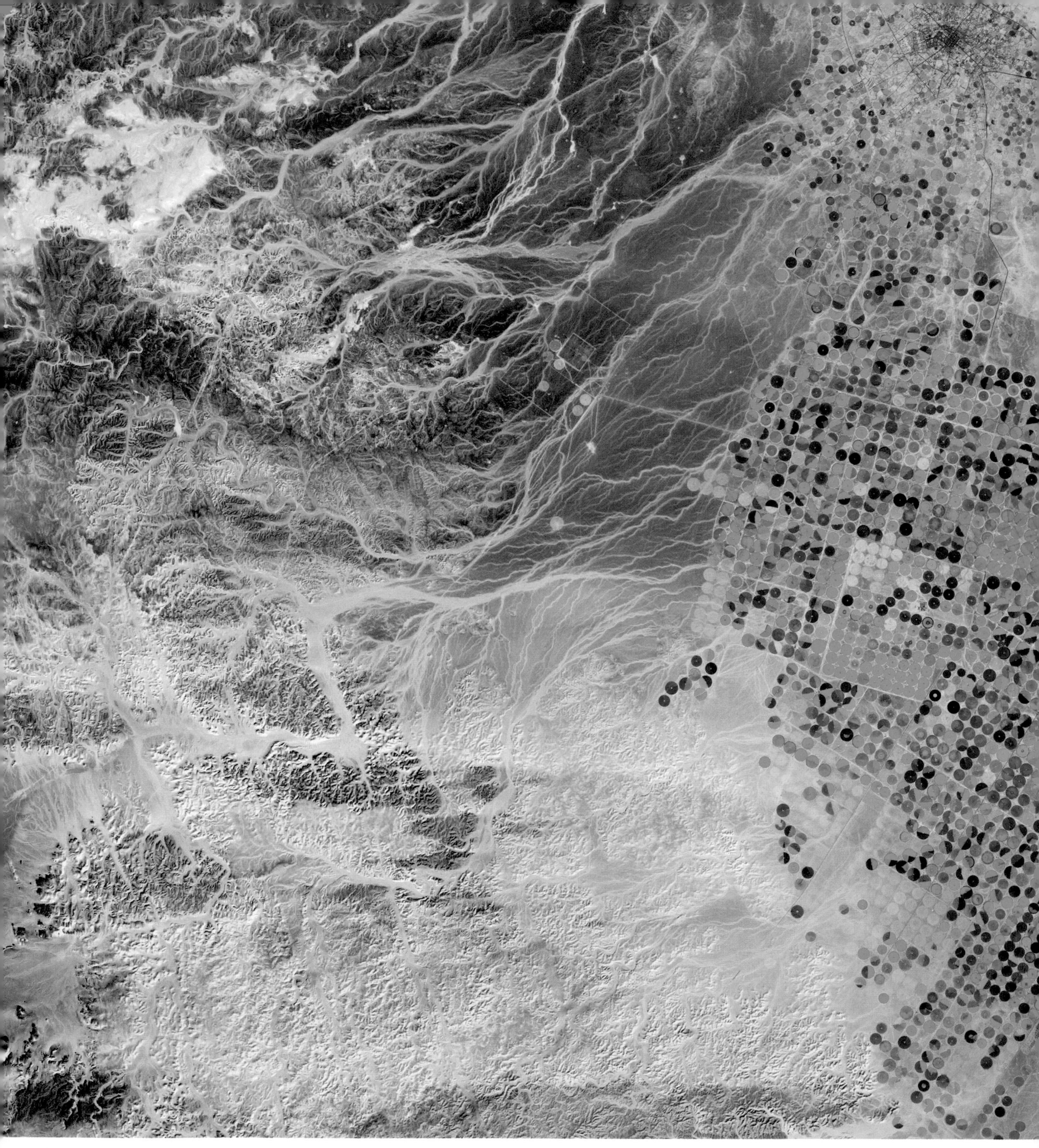

CONTROLLING WATER –

use of water resources for the generation of power and for the irrigation of otherwise infertile land for agriculture

Agricultural fields, Wadi as-Sirhan, Saudi Arabia

CONTROLLING WATER

Only a handful of truly wild rivers remain in the world – mostly in the Arctic and in a few remote rainforests. Virtually all the rest are tamed by dams and dykes which aim to prevent floods, irrigate crops, fill taps or generate electricity.

Since the construction of the 220 m-(722 foot-high) high Hoover Dam on the Colorado river in the USA in the 1930s, engineers have built hundreds of giant dams worldwide. These generate one-fifth of the world's electricity. China's Three Gorges Dam on the Yangtze is the largest yet, generating as much power as twenty coal or nuclear power stations.

Highest dams

		Height	
	Country	m	feet
Rogun	Tajikistan	335	1 100
Nurek	Tajikistan	300	984
Xiaowan (Yunnan Gorge)	China	292	958
Grand Dixance	Switzerland	285	935
Inguri	Georgia	272	892
Vaiont	Italy	262	860
Manuel M. Torres	Mexico	261	856
Tehri	India	261	856
Álvaro Obregón	Mexico	260	853
Mauvoisin	Switzerland	250	820

Largest volume embankment dams

		Volume (thousands)	
	Country	cubic m	cubic feet
Tarbela	Pakistan	148 500	194 238
Fort Peck	USA	96 050	125 633
Tucuruí	Brazil	85 200	111 442
Atatürk	Turkey	85 000	111 180
Yacyretá	Argentina	81 000	105 948
Rogun	Tajikistan	75 500	98 754
Oahe	USA	70 339	92 003
Guri (Raúl Leoni)	Venezuela	70 000	91 560
Parambikulam	India	69 165	90 468
High Island West	China (Hong Kong)	67 000	87 636

Major hydro-electricity plants

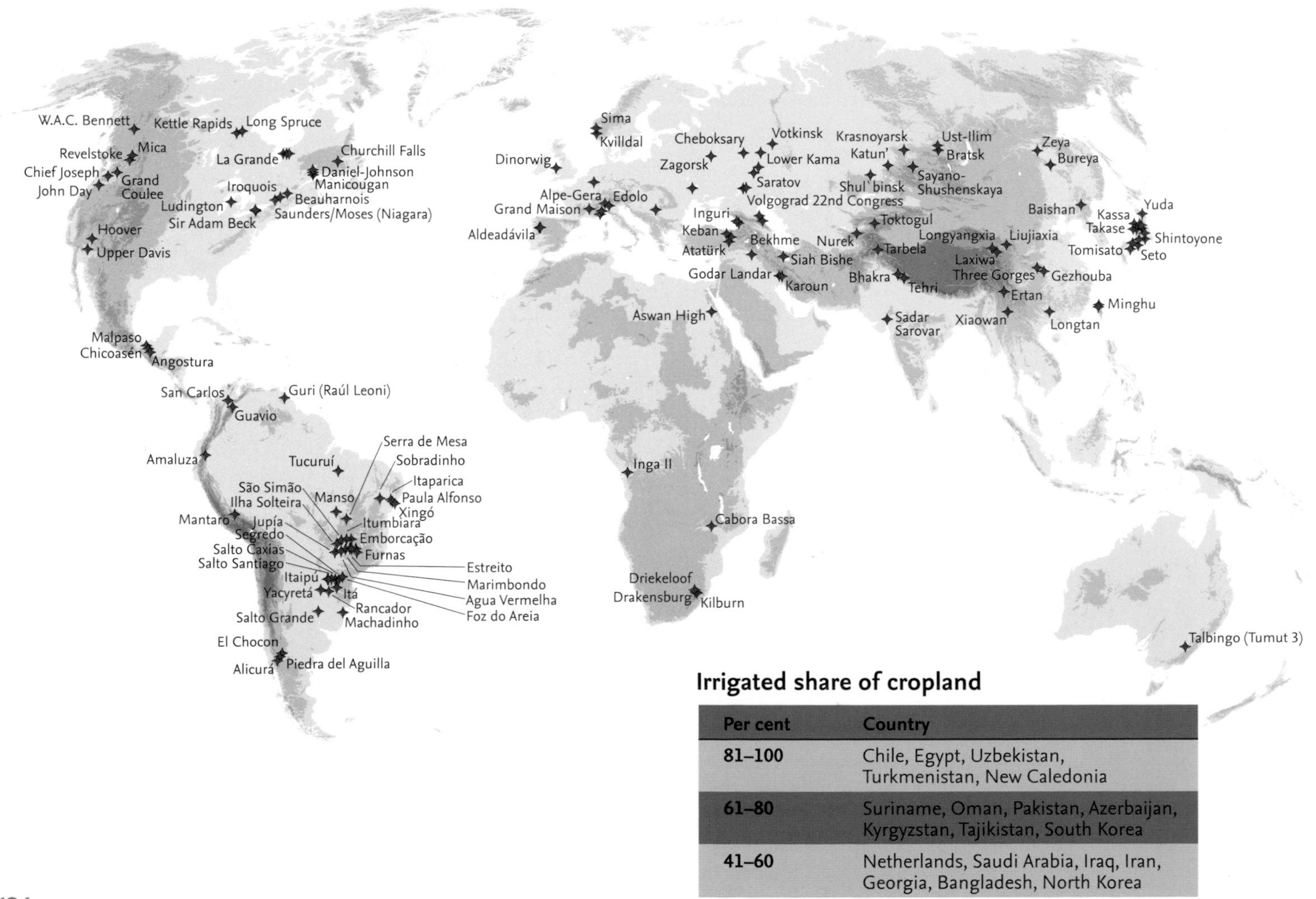

Irrigated share of cropland

Per cent	Country
81–100	Chile, Egypt, Uzbekistan, Turkmenistan, New Caledonia
61–80	Suriname, Oman, Pakistan, Azerbaijan, Kyrgyzstan, Tajikistan, South Korea
41–60	Netherlands, Saudi Arabia, Iraq, Iran, Georgia, Bangladesh, North Korea

Largest capacity reservoirs by volume

	Country	Capacity (millions) cubic m	Capacity (millions) cubic yards
Lake Kariba	Zimbabwe/Zambia	180 600	236 225
Bratsk Reservoir	Russian Federation	169 000	221 052
Lake Nasser (Aswan High dam)	Egypt	162 000	211 896
Lake Volta (Akosombo dam)	Ghana	147 960	193 531
Manicouagan Reservoir (Daniel-Johnson dam)	Canada	141 851	185 541
Lake Guri	Venezuela	135 000	176 580
Williston Lake (W.A.C. Bennett dam)	Canada	74 300	97 184
Krasnoyarsk Reservoir	Russian Federation	73 000	95 484
Zeya Reservoir	Russian Federation	68 400	89 467
Robert-Bourassa Reservoir (La Grande 2 dam)	Canada	61 715	80 723

Largest hydroelectric plants

	Country	Planned generating capacity (MW)
Three Gorges Dam (Sanxia)	China	18 200
Itaipú	Brazil/Paraguay	14 000
Guri (Raúl Leoni)	Venezuela	10 200
Tucuruí	Brazil	8 370
Sayano-Shushenskaya	Russian Federation	6 400
Grand Coulee	USA	6 180
Krasnoyarsk	Russian Federation	6 000
Robert-Bourassa	Canada	5 600
Churchill Falls	Canada	5 400
Longtan	China	4 700

Meanwhile, large dams have allowed governments to triple the amount of water delivered to farmers to irrigate crops in the past forty years, enabling the world to feed its fast-growing population. More than two-thirds of all the water used by humans supplies irrigated fields which grow almost 40 per cent of the world's food – and more than 80 per cent in countries such as Egypt and Turkmenistan.

But there is a big downside. In 2000, experts on the World Commission on Dams reported that many dams around the world have caused environmental and social havoc which outweighs the economic gains. Dams have flooded fertile valleys and forests, created millions of refugees, destroyed fisheries, emptied wetlands, eroded river banks and caused flooding.

Today, overuse of water has left big rivers like the Indus in Pakistan and the Colorado river in the USA and Mexico regularly running dry. And many farmers are turning to underground water. But while the rains eventually replenish river flows, many underground water reserves are being emptied forever. Water tables are falling fast as countries such as Saudi Arabia and India pump up far more water than is replenished by rains. The world is growing ever more of its food with water which is not being replaced.

Three Gorges Dam project, China

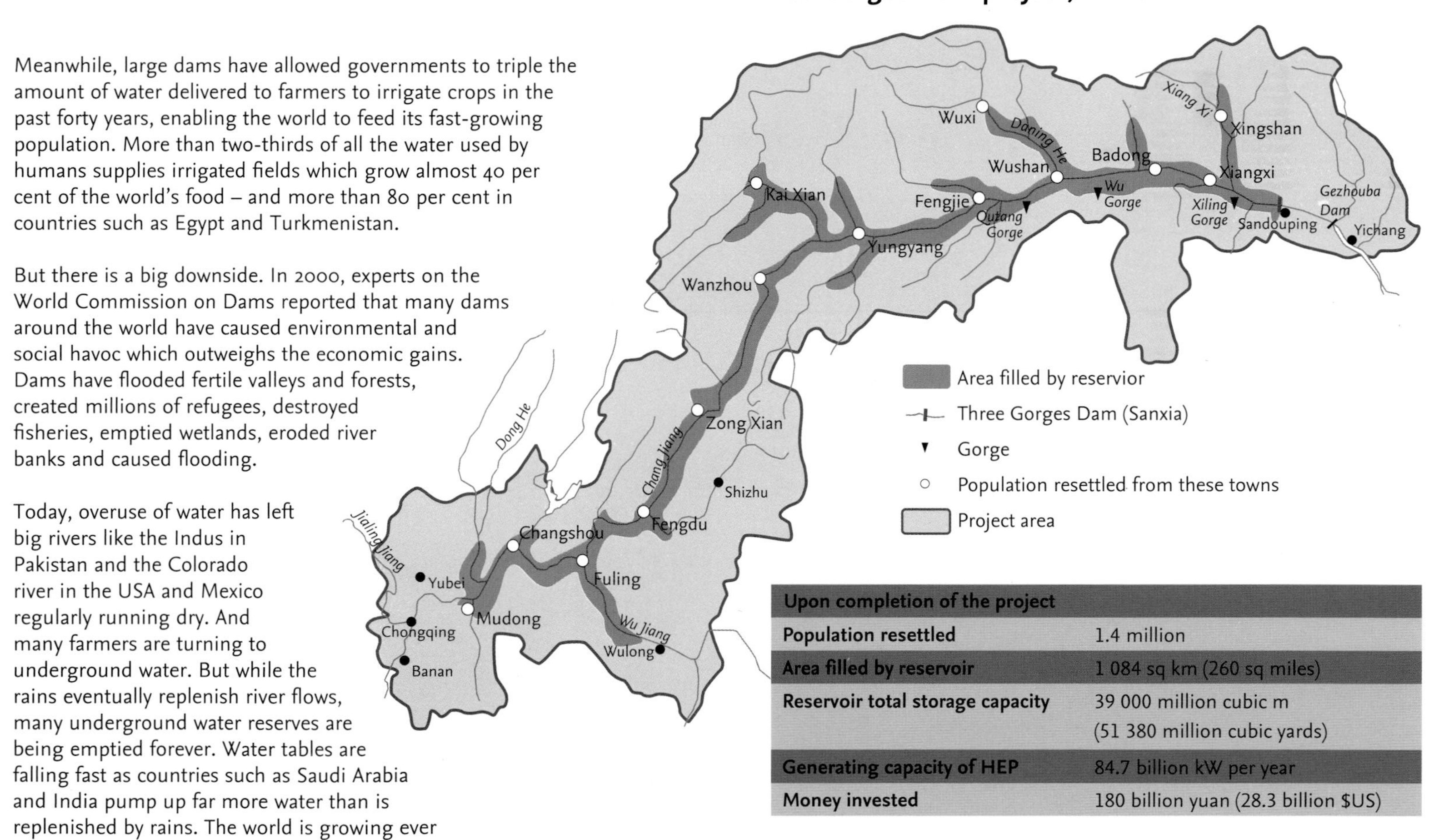

Upon completion of the project	
Population resettled	1.4 million
Area filled by reservoir	1 084 sq km (260 sq miles)
Reservoir total storage capacity	39 000 million cubic m (51 380 million cubic yards)
Generating capacity of HEP	84.7 billion kW per year
Money invested	180 billion yuan (28.3 billion $US)

The Yangtze is seasonal river. Many millions of Chinese live in cities downstream from the dam, where there are also important industrial areas and farming land. Control of flooding is an important part of the function of the dam. With a flood storage capacity of 22 000 million cubic m (28 800 million cubic yards) it has already shown it has the capacity to capture floodwater and limit the river flow downstream. The dam also discharges the reservoir each year during the dry season – December to March – benefiting shipping, agriculture and industry.

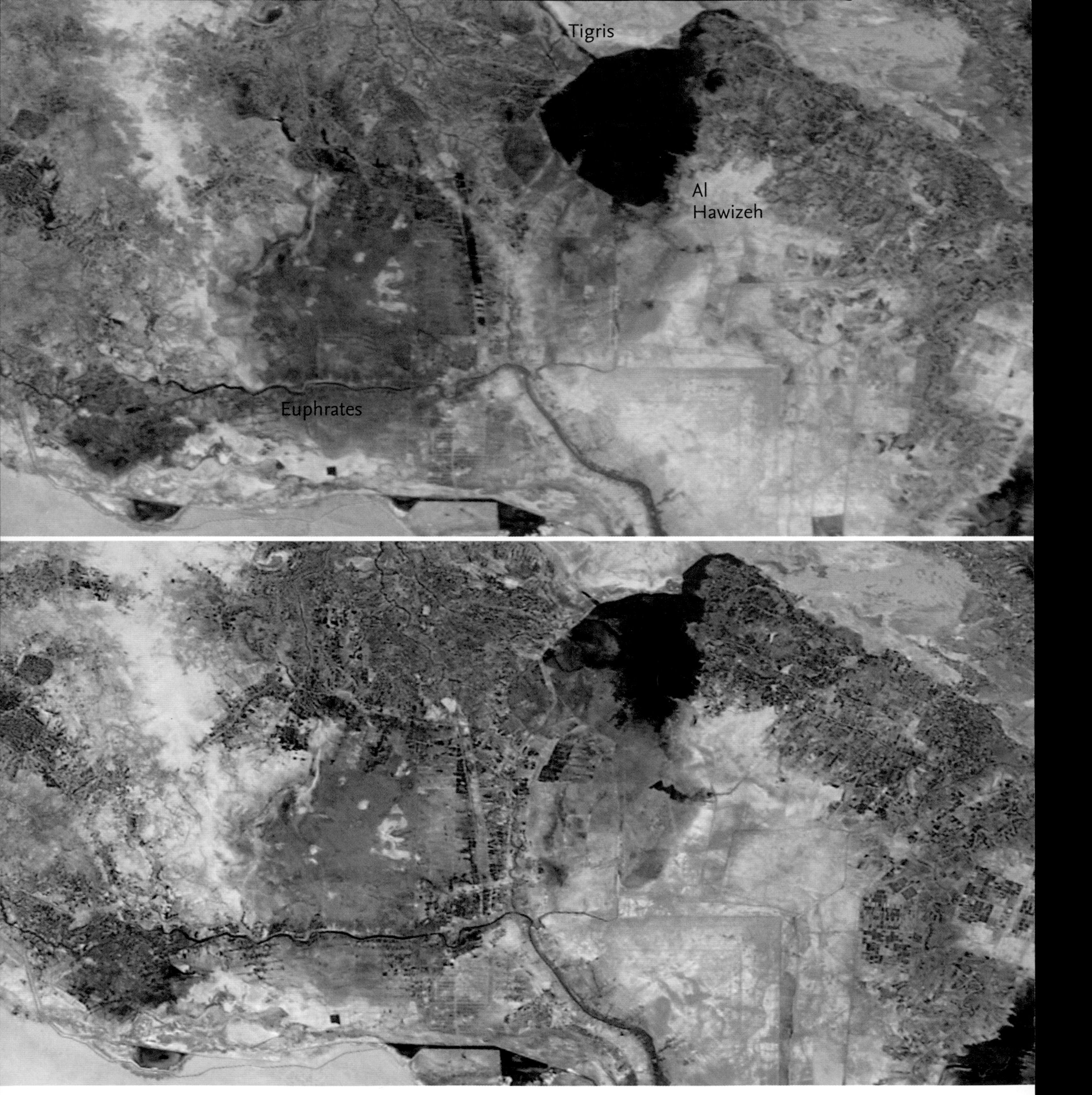

MESOPOTAMIA MARSHES, IRAQ 2000 AND 2003

The wetlands of Mesopotamia were once an area diverse in plant and animal life, a rare aquatic landscape in the desert. Over time the volume of water reaching them has been much reduced by dams to control floods, canals, reservoirs and hydroelectric schemes. In the 1990s they were drained by former Iraqi leader Saddam Hussein, partly as a punishment to the local tribes who had taken part in anti-government rebellions. In 2000 there is only one large remnant of marsh, Al Hawizeh, and this was to shrink further as spring floods no longer reached it.

MESOPOTAMIA MARSHES, IRAQ 2007 AND 2010

After the end of the Second Gulf War in 2003, many of the canals draining the marsh were demolished, several large marsh areas flooded again, Al Hawizeh expanded and irrigated crops were grown as can be seen in 2007. A drought in 2009 had a severe impact on the area, affecting crops, but in 2010 while some marshes have shrunk, irrigated crops seem to be extensive and healthy. There are still problems to be faced, the marshes are not interconnected as they once were and the incoming water levels may fall again as more extensive agriculture increases elsewhere.

FLOOD DEFENCES IN SOUTH AND NORTH HOLLAND, NETHERLANDS 24 SEPTEMBER 2002

The Netherlands is a low-lying country, where 25 per cent of the land surface lies below sea level and a further 50 per cent is less than a metre above sea level. The low elevation leaves the country vulnerable to large-scale flooding. In 1953, over 1800 people drowned following the collapse of several dykes that led to areas being inundated with floodwater. The simulated natural-colour satellite image is of the area southwest of Rotterdam; the intricate system of canals, dams, bridges, dykes and locks, are all designed to reduce the risk of flooding.

MERORWE DAM, SUDAN 5 OCTOBER 2010

The Merowe Dam is the largest contemporary hydropower project in Africa. It is located on the Nile's fourth cataract, 350 km (218 miles) north of Khartoum. The dam was built between 2003 and 2009, is 9.3 km (6 miles) long and up to 67 m (220 feet) high. Its capacity is 1250 megawatts and it doubled Sudan's electricity generation, enabling further socio-economic development in the country. However, creating the reservoir, which extends 174 km (108 miles) upstream, displaced over 50 000 people who were forced to move from the fertile Nile Valley to arid desert sites.

EXPANDING CITIES –

outward and upward growth of urban areas largely in response to the migration of people from rural areas

Burj Khalifa skyscraper, Dubai, UAE

EXPANDING CITIES

Humanity is becoming a predominantly urban species. Cities occupy just 2 per cent of the Earth's land surface, but they now house over half of the world's population. And, perhaps most alarming of all, they use three-quarters of the resources we take from the Earth. The growth of the world's largest cities in particular has been staggering. The first megacity – with a population of ten million people – was New York, which reached that figure around 1940. Today there are twenty-two megacities, half of them in Asia, including three in India and two in China. The largest, Tōkyō in Japan, has over thirty-five million inhabitants. Most urban growth is from migration. China, which has ninety cities with populations of more than a million people, has a 'floating population' of more than 100 million people who have moved to cities in recent years. Its cities expect to welcome another 400 million people from the countryside within the next thirty years.

Cities have become economic powerhouses, providing jobs in manufacturing, transport and wholesaling, as well as services including universities, hospitals, government services, banking, media and culture. Saõ Paulo contributes 40 per cent of Brazil's GDP; the Shanghai area takes a similar percentage of foreign investment in China. But, despite their attractions, cities can be centres of squalor too. More than a quarter of all urban inhabitants live in unplanned, overcrowded and often illegal squatter settlements, with no running water, let alone tickets to the opera. And cities can become victims of their size. Congestion, worsening air pollution and crime can cause megacities to stop growing, as people and businesses flee to the suburbs or surrounding new cities. Smoggy Mexico City had sixteen million people in 1984 and was widely expected to grow to thirty million by 2000; but instead it has stopped growing and fluctuates around twenty million. The result of this out-migration has been a new geographical phenomenon known as the 'polycentric megacity zone' – an urban landscape composed of a number of different centres. These new zones include the Yangtze delta region around Shanghai, southeast England around London, and the Japanese urban corridor between Tōkyō and Ōsaka. Better electronic communications are encouraging this trend.

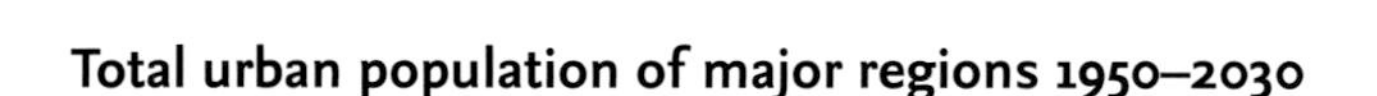

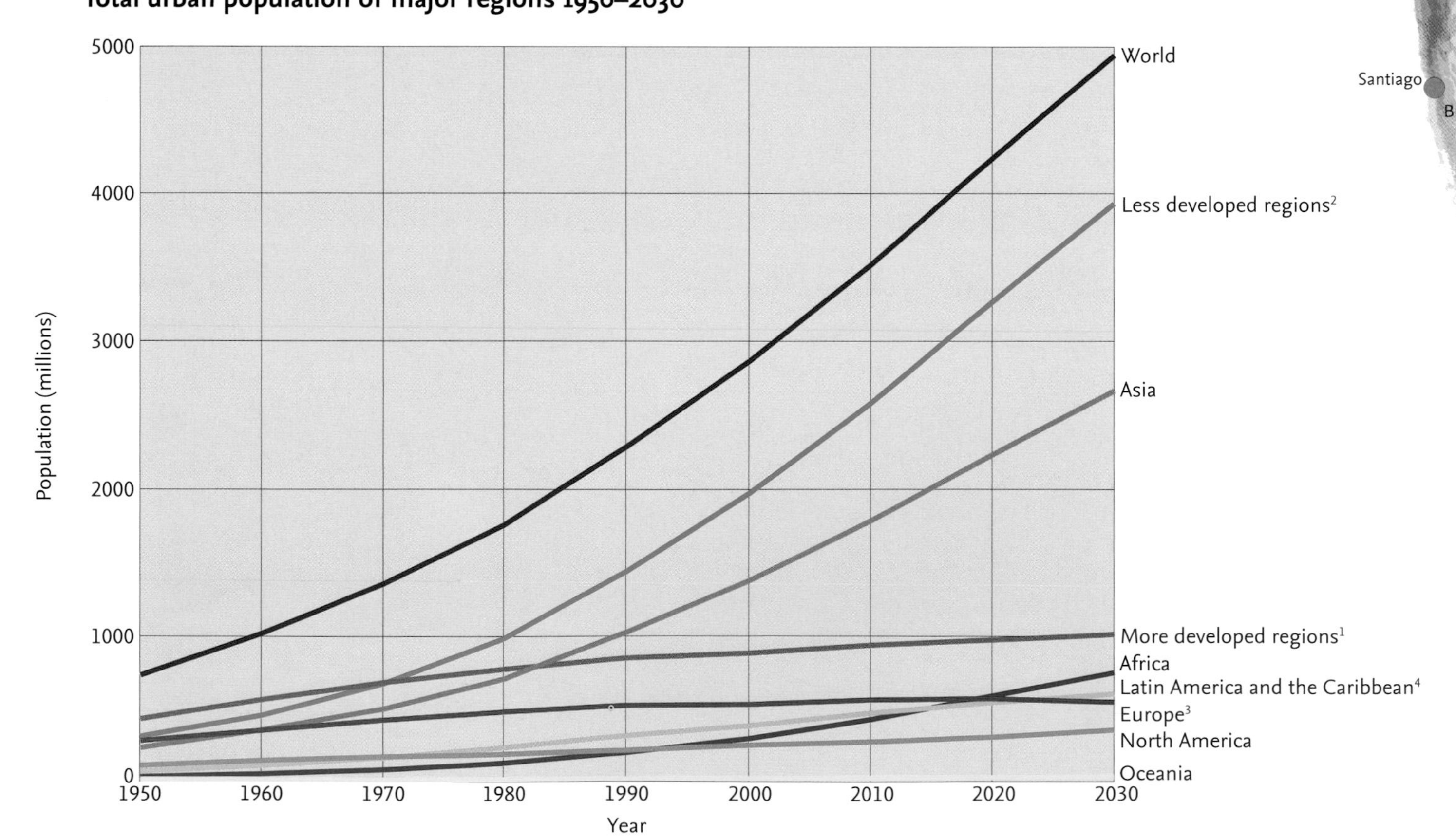

1. Europe, North America, Australia, New Zealand and Japan.
2. Africa, Asia (excluding Japan), Latin America and the Caribbean, and Oceania (excluding Australia and New Zealand).
3. Includes Russian Federation.
4. South America, Central America (including Mexico) and all Caribbean Islands.

The World's major cities

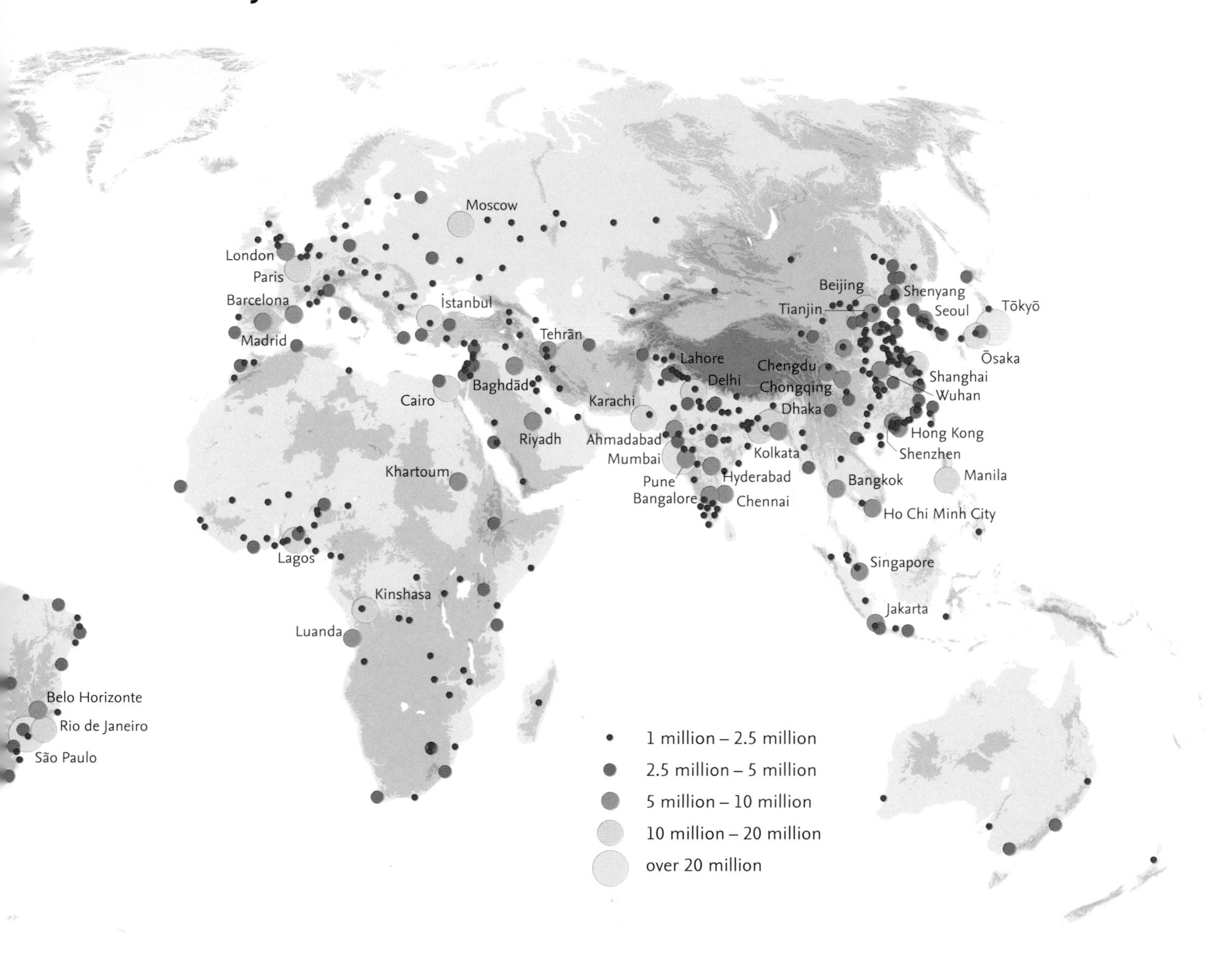

World's largest urban agglomerations

1955

Agglomeration	Country	Population (millions)
Tōkyō	Japan	13.713
New York	USA	13.219
London	UK	8.278
Shanghai	China	6.865
Paris	France	6.277
Buenos Aires	Argentina	5.843
Rhein-Ruhr	Germany	5.823
Moscow	Russian Federation	5.749
Chicago	USA	5.565
Los Angeles	USA	5.154

2005

Agglomeration	Country	Population (millions)
Tōkyō	Japan	35.327
Mexico City	Mexico	19.013
New York	USA	18.498
Mumbai (Bombay)	India	18.336
São Paulo	Brazil	18.333
Delhi	India	15.334
Kolkata	India	14.299
Buenos Aires	Argentina	13.349
Jakarta	Indonesia	13.194
Shanghai	China	12.665

2015 (projected)

Agglomeration	Country	Population (millions)
Tōkyō	Japan	37.049
Delhi	India	24.797
Mumbai (Bombay)	India	21.199
São Paulo	Brazil	21.300
Mexico City	Mexico	20.078
New York	USA	19.968
Shanghai	China	17.840
Kolkata	India	16.924
Dhaka	Bangladesh	16.623
Karachi	Pakistan	14.818

The growth of cities with over 5 million inhabitants

Urban sprawl itself is emerging as an environmental menace. Concrete and asphalt cover rich farmland around fast-expanding megacities such as the Indonesian capital Jakarta. And sprawl encourages ever greater use of motor vehicles, which cause smog and emit the gases behind climate change.

Cities have very large ecological 'footprints'. To service Londoners, and absorb the pollution they create, requires an area of land 120 times the size of the city itself. Sprawling cities such as Las Vegas have even bigger footprints. Many of the resources cities require may come from distant lands, but cities still put huge strains on local resources such as water. Air pollution from Chinese cities cuts crop yields across the country by up to one-third.

Yet some say that cities may be the key to a more sustainable future. In developing countries, people who move to cities from the countryside have fewer children. Urban children can be seen as a burden, requiring education for instance, rather than as a useful source of farm labour. So urbanization may be the ultimate solution to population growth.

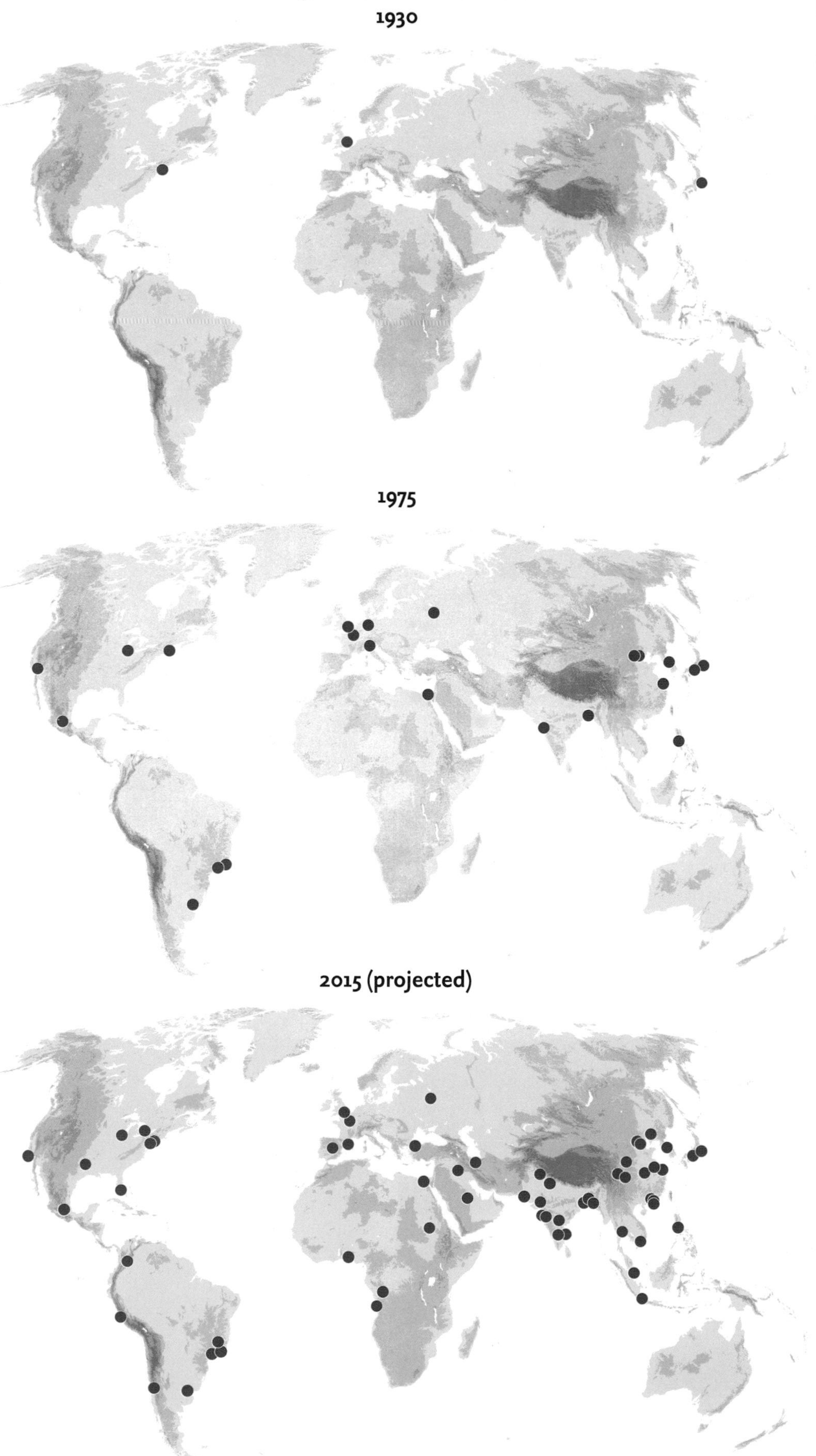

Level of urbanization by major region 1970–2030

Urban population as a percentage of total population

	1970	2010	2030
World	36.1	50.5	59.0
More developed regions[1]	64.7	75.2	80.9
Less developed regions[2]	25.3	45.1	55.0
Africa	23.6	40.0	49.9
Asia	22.7	42.2	52.8
Europe[3]	62.8	72.8	78.4
Latin America and the Caribbean[4]	57.1	79.6	84.9
North America	73.8	82.1	86.7
Oceania	70.8	70.2	71.4

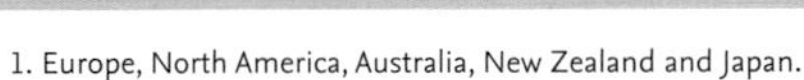
1. Europe, North America, Australia, New Zealand and Japan.

2. Africa, Asia (excluding Japan), Latin America and the Caribbean, and Oceania (excluding Australia and New Zealand).

3. Includes Russian Federation.

4. South America, Central America (including Mexico) and all Caribbean Islands.

The inhabitants of cities recycle more of their waste, and use public transport more often than their country cousins. They even grow food. The proximity of millions of customers means every scrap of spare land is cultivated. Around 15 per cent of the world's food is grown within city limits. Shanghai still produces most of its own milk, eggs and vegetables. Amazingly too, cities can be centres of wildlife. One small derelict industrial site in London has 300 species of plants growing – many times more than the countryside around the city. Cities may still be global parasites but they have redeeming features.

Tallest buildings 1975 and 2012

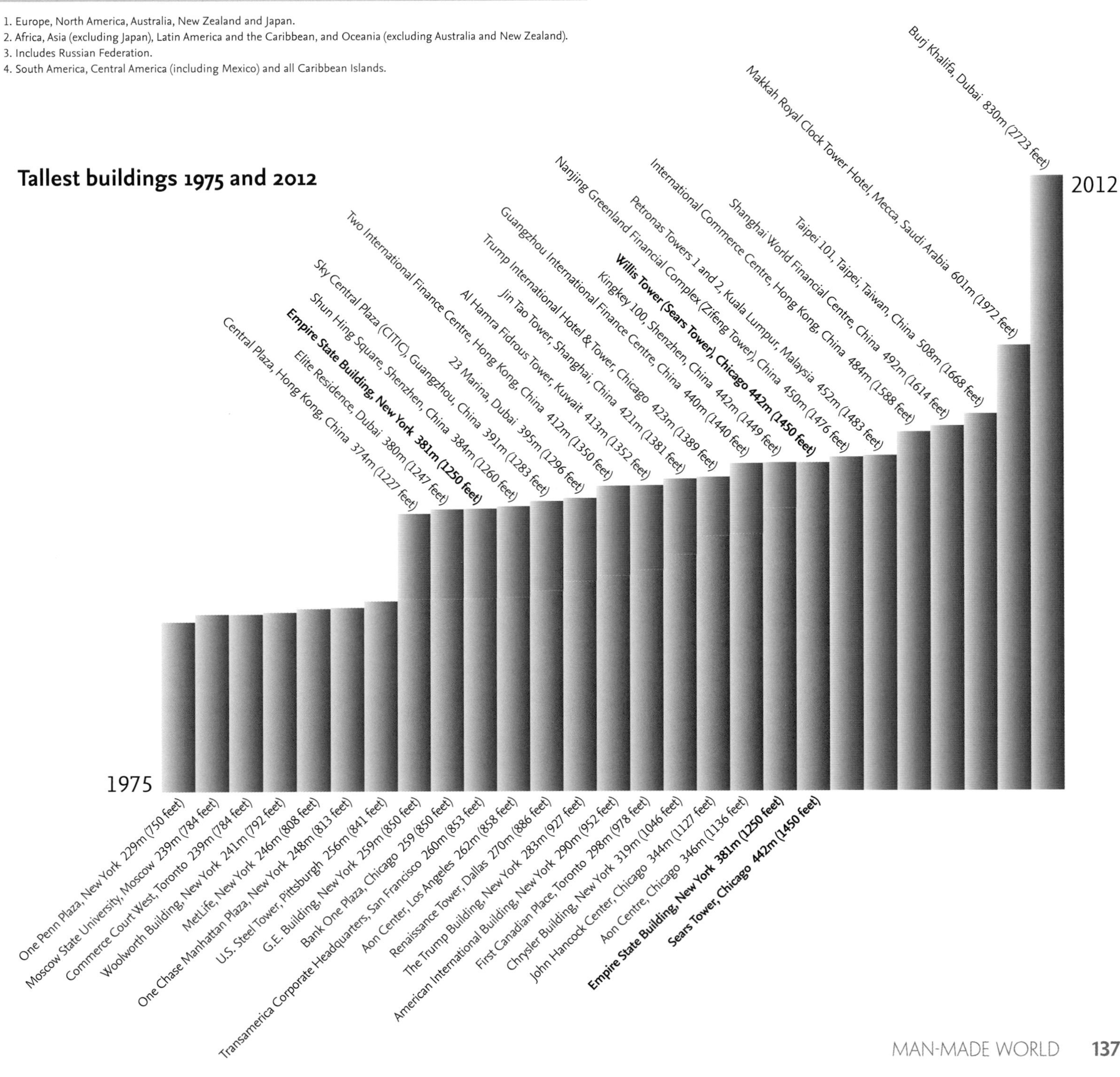

ABERDEEN HARBOUR, HONG KONG 1920

Taken in 1920, this photograph of Aberdeen Harbour, Hong Kong shows an anchorage full of the traditional Chinese sailing vessels, junks.

The surrounding land contains few buildings although in the distance some 'low rise' buildings are visible. Contrast that with the recent view opposite.

LOST TRANQUILITY 2005

The entire harbour area, which has now been reduced in size, is enclosed by multi-storey apartment blocks built to accommodate the ever-increasing population. When land is at a premium, reclaimed areas have to be put to intensive use by building higher.

SINGAPORE 1962

When this photograph of Singapore was taken, the multi-storey Asia Insurance building to the left was the tallest in the city and the harbour was used by traditional sampans for trading and fishing in the coastal waters around the island of Singapore.

HIGH-RISE DEVELOPMENT 2010

Although the Asia Insurance building still exists, it is no longer the tallest in the city. It is now dwarfed by many multi-storey commercial buildings which have sprung up in the last fifty years in the central area of the city as Singapore has expanded as a major commercial centre. Even away from the business district much of the land of Singapore island is built up with the majority of Singaporeans living in high-rise flats. The growth of the city has led to land reclamation for the airport and for the expansion of the port.

TEHRAN, IRAN 2 AUGUST 1985…

Tehran, Iran's capital, is one of the world's fastest-growing cities. It lies on the slopes of the Elburz Mountains with archaeological evidence dating from 6000 BC. Tehran was originally a village outside the ancient city of Ray. When Ray was obliterated in the thirteenth century during the Mongol invasion, Tehran became inundated with refugees. Tehran's importance and population continued to grow, and it became Iran's thirty-second capital city in 1795. In the 1940s, its population was 700 000; in 1966 it was three million, and by 1986 it had doubled to six million.

… AND 19 JULY 2009

The two false-colour satellite images show the city's rapid expansion over a relatively short period of time. In both images, barren areas are brown, vegetation appears bright green, and urban areas are grey or black. By 2011 Tehran was the world's thirty-seventh largest city, with a population of almost eight million. Consequences of urbanization include loss of arable land, and air and water pollution. It is undecided whether Tehran will remain Iran's capital city; their Government is considering relocating the capital to a city which is less at risk from major earthquakes.

SHENZHEN, GUANGDONG PROVINCE, CHINA 15 NOVEMBER 1999...

Shenzhen, located on the Pearl River Delta in southern China's Guangdong Province, is one of the world's fastest growing cities. In the early 1970s, Shenzhen was still a small fishing village. However, in 1980 the area was chosen to be one of China's five Special Economic Zones, and the city began to expand rapidly due to foreign investment in the manufacturing and service industries. Numerous new factories were built and the new jobs attracted a huge level of migration from inland China that boosted the city's population and led to rapid urbanization.

… AND 1 JANUARY 2008

The two satellite images, taken less than ten years apart, show the rapid level of urbanization in Shenzhen. The city has built up along the coast, especially around the port, and also inland to the east of the areas of higher ground. Low-lying areas were developed first and, as population pressure increased, smaller hills were flattened to provide more land to build on. Shenzhen is the third busiest port in China after Shanghai and Hong Kong. Salt pans are still evident along the coast to the northwest of the city. In 2009, Shenzhen's population reached 9 827 000.

KATHMANDU, NEPAL 1969 AND TODAY

In the near forty-year period from 1969, when the top photograph was taken, the population of Kathmandu, the capital of Nepal, increased five-fold due mostly to migration from the countryside. This resulted in the built-up area of the city expanding outwards to swallow up rural communities. The lower photograph, taken from the same location, shows a dramatic change in the landscape in this relatively short time. The city continues to grow and it is estimated that the population will reach over 1.25 million by 2021.

SÃO PAULO, BRAZIL 2004

The shanty town of Favela Morumbi is one of São Paulo's largest slum areas and is typical of the unplanned expansion seen in many underdeveloped or developing countries as thousands of country dwellers flood into the main cites in search of a better life. Conditions are often poor, with little or no sanitation and in some cities such unofficial developments are razed to the ground by the authorities, only to spring up again.

IMAGINING THE FUTURE

Julia Bucknall

Man-made changes to our world are inevitable, and necessary as we seek ways to meet the needs of our steadily growing population. Over half of the world's population now live in urban areas and the rapid growth of cities – their physical expansion sometimes dependent on huge land reclamation projects – epitomize such changes and the great demands they place on our natural resources. Developments can often be unsympathetic to the scarcity, quality and variability of resources, and commonly harm our most precious resource – water. Most people reading this probably realize we have a water crisis and will have seen some version of the figures:

- 780 million people lack access to improved water, 2.4 billion people are without access to improved sanitation, half of whom defecate in the open. Result: 4 000 children dying unnecessarily from diarrhoea every day.

- Massively increasing demand for water for food. If we carry on producing food as we do now, we would need 45 per cent more water to produce enough food to feed the world in 2030 – water for which there will be many more calls in future.

- Greatly increasing demand for water in energy. Most of the new fuels, fuel extraction or production techniques, and most of the new energy generation techniques on which the world pins hopes for a green low carbon future, are much more water-intensive than the current fuel mix. For example, 90 per cent of the existing electricity generation capacity in developing countries is water-intensive. As electrification grows, demands for water for energy will increase. Energy accounts for about 20 per cent of consumptive water use in the US, compared to a global figure of about 10 per cent

- Floods and droughts. Every year 300 million people are affected by water-related catastrophes. And those figures will only increase as cities get larger (increasing the number of people and assets at risk) and as climate change accelerates the hydrological cycle.

- Environmental stress. Even now, we are living off our water nest egg by consuming year after year the stocks of water stored naturally underground. About 60 per cent of urban water, and 50 per cent of water used to irrigate agriculture, comes from groundwater, much of it extracted at rates that massively exceed the safe recharge rate. We are now bringing so much groundwater up into the hydrological cycle that we are causing sea levels to rise. Some of the world's mightiest rivers fail to reach the sea because we draw off so much water. This has a huge impact on pollution, soil quality, coastal ecosystems, and fisheries.

So what is the future we want and how do we get there? In my view we should envision five things:

- A world where no one's health, livelihood, physical security or ecosystems should be unreasonably held back by poor management of water. This means, of course, that everyone has safe water and basic sanitation services that are sustainable in a financial, operational, and environmental sense. Getting there means making political changes so that governments either set prices for water high enough to cover the costs of providing the service (while subsidizing people who cannot afford the charges) or transfer enough money predictably and transparently to water companies or rural systems. And, with the right incentives, operators will make huge strides to deliver clean water more efficiently. Otherwise, services stay in the negative spiral of poor services, budget overruns, deferred maintenance, worse service, and so on.

- A situation in which everyone involved in water management knows how much water exists in each basin, where and how efficiently it is used, and everyone uses the same definition of water-related terms. This sounds surprising, but in my view it is surprising how little we know about our water, and how many decisions have to be made in conditions of great uncertainty. In fact, it is monumentally difficult to measure water, especially when we consider that so many basins originate in remote mountains, when so many rivers cross sensitive international borders and when so much of our water is underground. And terms like water use or water efficiency can mean completely different things depending on who is talking.

- A world in which governments around the globe set limits for water consumption from all users, including environmental services, in each basin and transparently enforce those limits. And a world in which users use their allocation – but no more – efficiently. Without those limits, even if they are super-efficient, users may continue using more water than is safe for the environment or fair to other users. In that respect water is different from energy.

- A world in which poor people, environmental advocates, industry, agriculture, and cities all have equal voice in the bargaining process when deciding how much water to allocate to whom.

- A world in which entrepreneurs respond to clear, transparent decision making, consistent enforcement of rules, and a predictable flow of revenues, by inventing solutions that will lead to dramatic changes in what is possible. We already see computer programmers, some amateur, coming up with software applications to help poor women in rural Africa ensure their water pump is mended. More and more towns and cities are desalinating water with renewable energy, constructing buildings to capture rainwater and reuse water. And scientists from NASA are able to show us the condition of water in the most remote areas and in the minutest detail.

This vision points to some priorities for moving forward. Investing in data to know our resources and share that knowledge, in processes and governance arrangements to make allocation, policy, and investment processes transparent, and reflecting a balance of the largest number of needs will be essential. Once the rules are clear, entrepreneurialism will lead to innovations we cannot yet imagine. We cannot construct our way to a water solution, even when large infrastructure is an important part of the solution. But we have the best chance of success if we tackle the challenge head on, make decisions based on evidence, set and enforce standards consistently, and allow the human spirit to find creative solutions to a problem so essential to our wellbeing and livelihoods.

Children collecting water at the end of the day, Ghana.

DAMAGED

WORLD

MAN'S IMPACT

Deforestation – the felling of trees, often illegally and in an unmanaged way, commonly to provide land for agriculture or industry

Pollution – the release into the environment of harmful or poisonous substances, often by-products of industrial or agricultural activity

DEFORESTATION – the felling of trees, often illegally and in an unmanaged way, commonly to provide land for agriculture or industry

Wildfire in South Africa

DEFORESTATION

Forests are the planet's largest, most biologically diverse and most critical ecosystems. They once covered some two-thirds of the Earth's land area, and still cover approximately a third, 40.3 million sq km (15.6 million sq miles). They stretch from the tropics to the edges of the Arctic tundra, and contain around half the world's plant and animal species. They play vital roles in recycling water around the planet, protecting soils and maintaining the natural carbon cycle. Mankind has always lived in, planted and destroyed forests. There are probably few true virgin forests anywhere. But the scale and pace of deforestation in the past 200 years dwarfs anything seen before. Some 99 per cent of Europe's 'old growth' forests outside Russian Federation are gone, along with 95 per cent of those in the continental USA. They have been either converted to farmland or urban areas or have been replaced with commercial plantations. Larger stands remain in the 'boreal' forests of northern Russian Federation, Canada and Alaska, and in the equatorial rainforests of the Amazon, central Africa and southeast Asia. But many of these forests are now disappearing fast, and could be gone within a decade.

Distribution of forests by region 2010

Region	Forest area sq km	Forest area sq miles	% of global forest area	% of region's land area
World	40 330 600	15 571 725	100.0	31
Africa	6 744 190	2 603 945	16.7	23
Asia	5 925 120	2 287 701	14.7	19
Europe	10 050 010	3 880 329	24.9	45
North and Central America	7 053 930	2 723 536	17.5	33
Oceania	1 913 840	738 937	4.8	23
South America	8 643 510	3 337 276	21.4	49

Annual net change in forest area by region 1990–2010

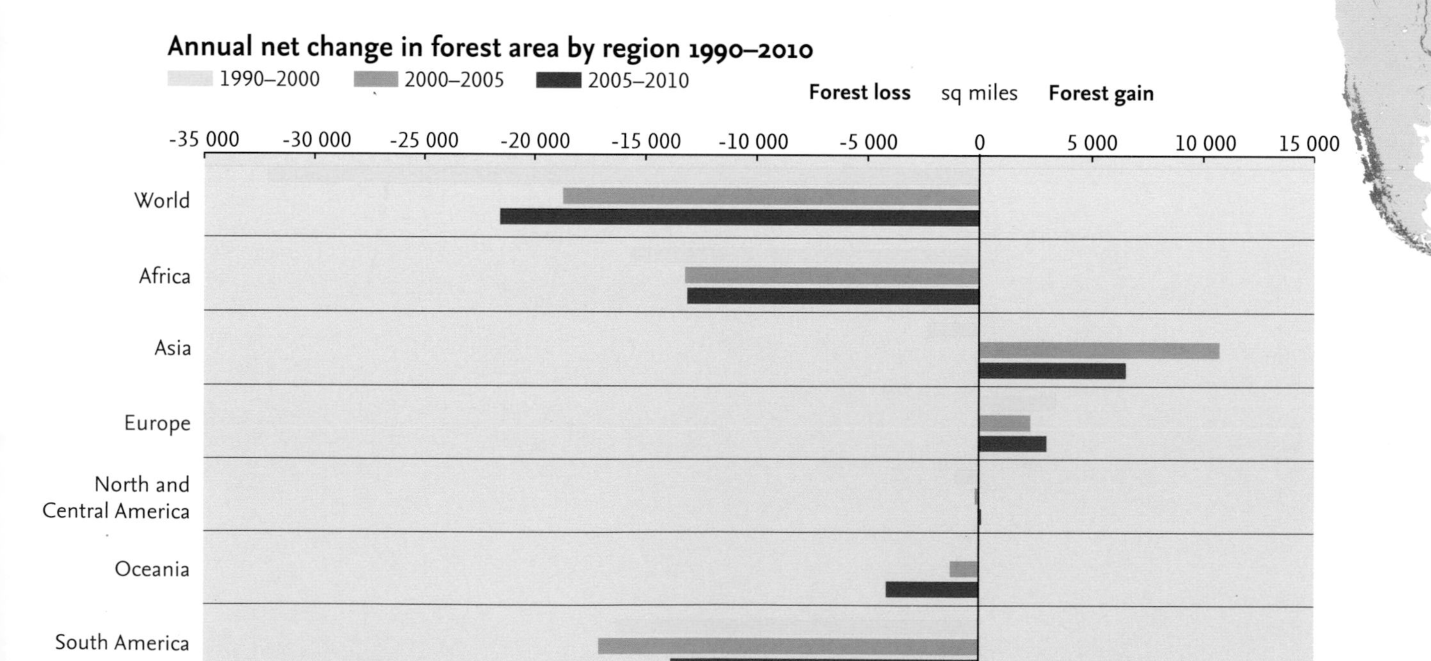

The world's forests

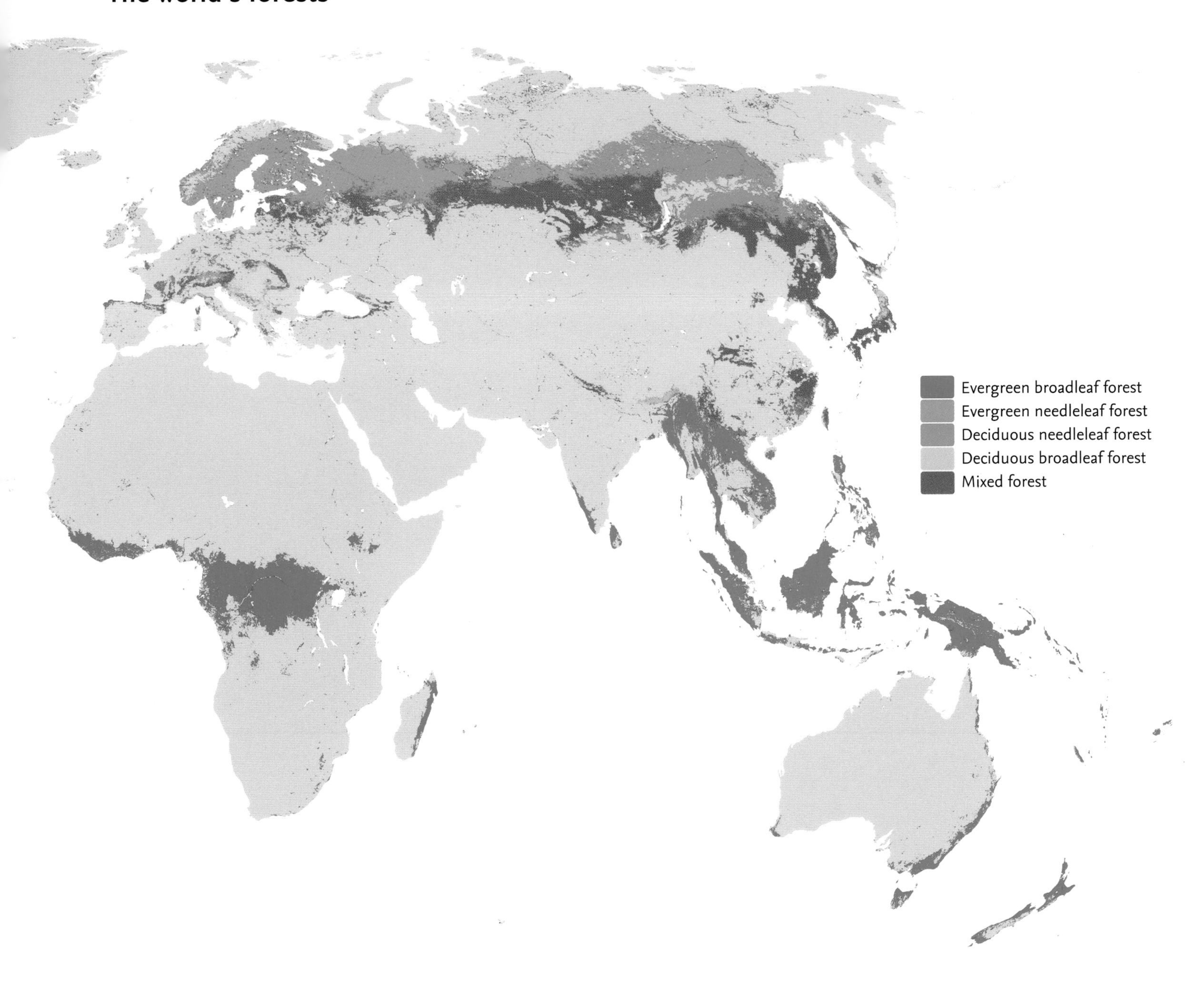

The pressures are greater in more densely populated countries, where migrant farmers burn forests to provide land for farming. But in the Amazon Basin, landowners are clearing forests to make way for cattle ranching and soya bean production, while logging is escalating in Africa. Indonesian islands such as Borneo and Sumatra suffer from both logging and clearance for palm oil plantations. Ironically, one growing market for palm oil is as an environmentally friendly 'biofuel' additive to vehicle fuels in Europe. The scale of deforestation is not the only issue, however. Piecemeal forest removal and penetration by roads break large forests into fragments which are less able to support a full range of species, because large animals have less room to roam and because fragmentation limits an ecosystem's ability to recover from catastrophes such as fires.

Lawlessness is a major problem in large, remote forests. The rights of indigenous communities to manage the forests are often abused. And conservation laws are widely flouted. An estimated 80 per cent of all logging in Indonesia is now illegal. In poor countries where forests are one of the few valuable natural resources, corruption in government and money-raising by armies, rebel groups and militias often contribute to forest destruction. Liberia, Myanmar (Burma) and Papua New Guinea are recent well-documented examples. Following a decision in 1998 to end logging of natural forests within its own borders, China is now the world's largest importer of tropical rainforest timber. But the timber, much of it illegally cut, ends up in products such as furniture which are sold in Europe, Japan and North America.

Largest annual net changes in forest area 2005–2010

Largest annual losses

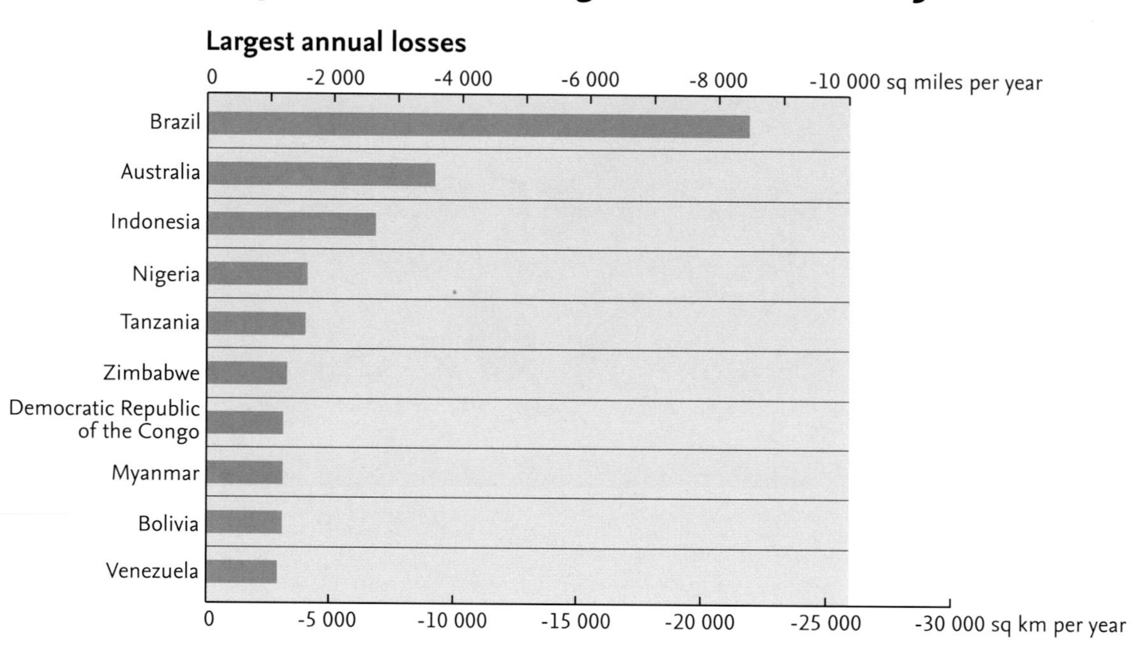

Largest annual gains

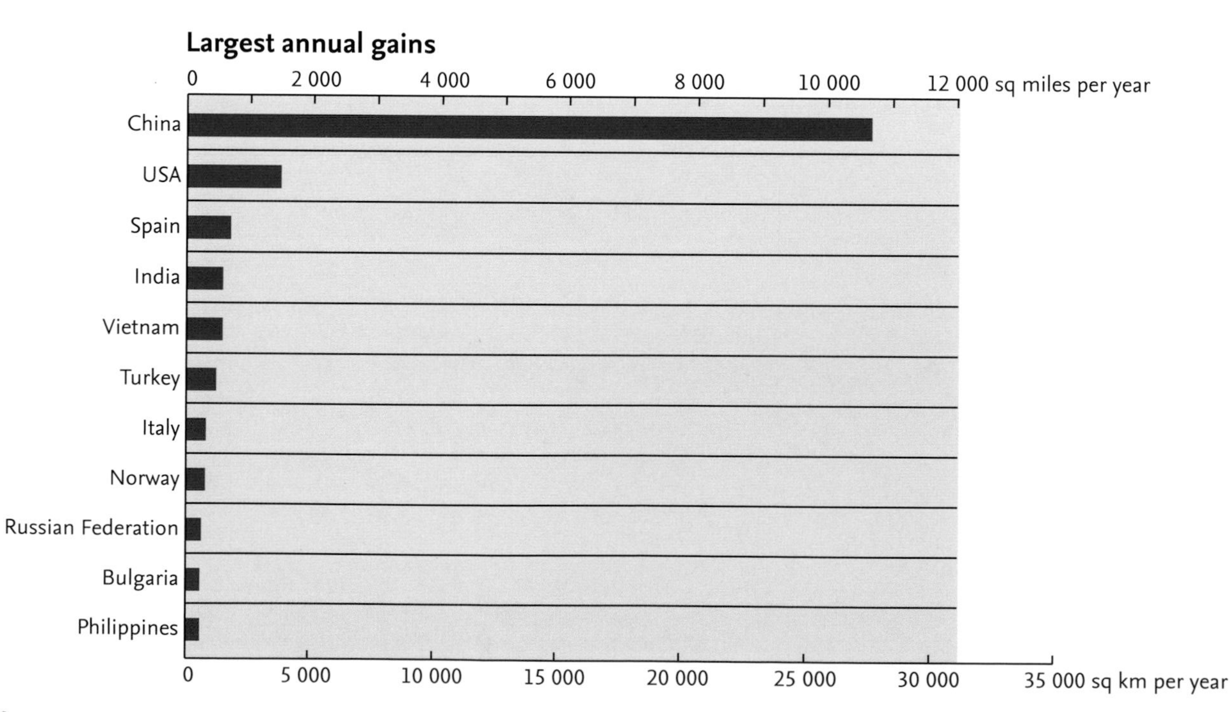

Deforestation in the Amazon Basin, Brazil

Countries with largest forest area 2010

Percentage of total world forest area

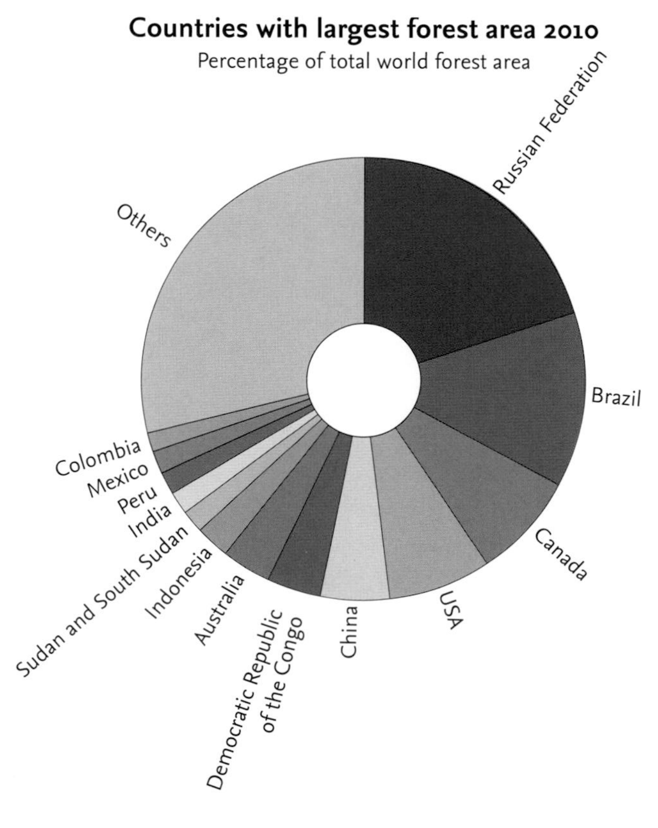

When forests are removed, soils often erode rapidly when exposed for the first time to direct rain. But in the longer run the destruction of forests may diminish rainfall. This is because, particularly in lush rainforests, trees collect and recycle rainfall back into the air through a process called evapotranspiration, keeping the winds moist and stimulating rainfall downwind. Remove the trees and the rains fail. The loss of rainforests in West Africa is thought to be one reason for the spreading Sahara desert. Deforestation can also alter global climates indirectly. Decaying timber eventually releases into the atmosphere carbon dioxide, a greenhouse gas which accelerates global warming. The loss of 1 sq km (0.4 sq miles) of forest will typically release 10 000 tonnes of carbon. Up to 20 per cent of current global warming may be due to carbon released from deforestation. In the past two decades, the temperate forests of Europe and North America have undergone modest increases in extent, thanks to the planting of commercial stands. Across the world, more forests are being declared national parks and reserves, but conservationists say many of these 'paper parks' offer little protection to surviving natural forests.

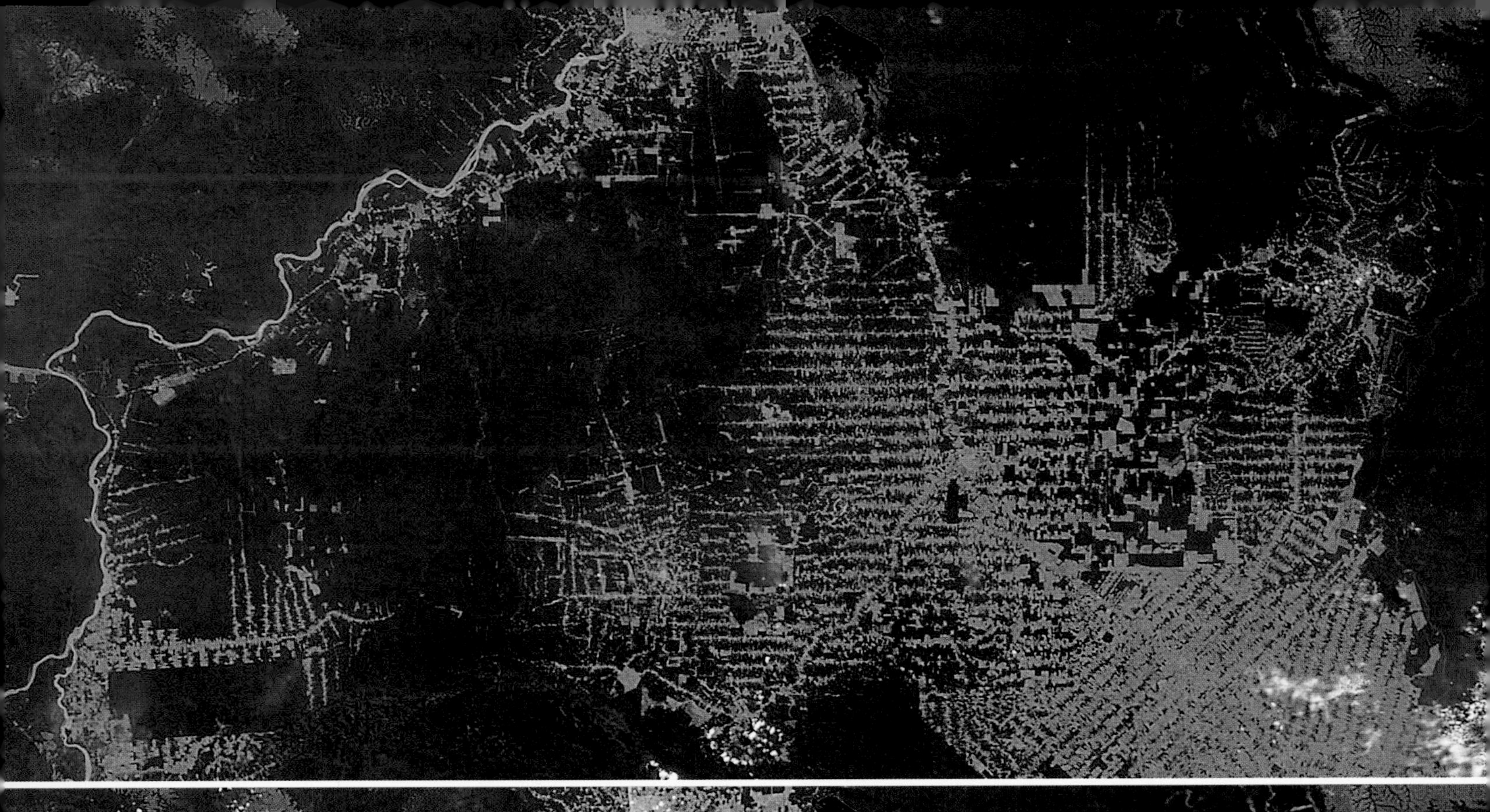

AMAZON DEFORESTATION, RONDÔNIA, BRAZIL 2000 AND 2003

Rondônia is one of the parts of the Amazon most affected by deforestation. The areas cleared of forest show up in the satellite images as tan, while crops and pasture show as light green. In this area the cleared forest moves both northwest and northeast along the roads. By 2003 it was estimated that nearly 68 000 sq km (over 26 000 sq miles) of the forest in the state had been cleared. This represents around one third of all the forest in the state. Over time a pattern emerges in the images. Early clearings look like fishbones extending out from the roads. These become a mix of cleared

AMAZON DEFORESTATION, RONDÔNIA, BRAZIL 2006 AND 2010

areas, settlements and some uncleared areas. Not all of the roads are legal, but small farmers still migrate to these areas, clear land for crops and when after a short time the rain and erosion affect the soil and the crop yields fall away, they use it for cattle pasture and move to another area to clear. When they have cleared all the land they claimed it often ends up merged with other claims as large areas of cattle pasture.

FOREST CLEARED FOR CATTLE AND PLANTATIONS, MATO GROSSO, BRAZIL 8 AUGUST 2008 AND 2010

The images above show two areas in Mato Grosso in Brazil where rainforest has been cleared for different agricultural purposes – cattle farming and soybean plantations. During the past forty years approximately 20 per cent of the Amazon rainforest has been cut down; an estimated 70 per cent of the cleared land was used for cattle farming. Fortunately the rate of deforestation has gradually started to slow down over the last six years.

REFORESTATION, MATO GROSSO, BRAZIL 2009

The entire ecosystem of the Mato Grosso region in Brazil is threatened by the impacts of agriculture. Reforestation is a practice that has been used more in recent years to replace forest which has been lost. Reforested areas can be picked out from the air as the tree-line is often even and unbroken, with leaves the same shade of green and patterns can be seen within the tree plantation area. Projects involving multiple native species are being encouraged.

FORESTED, DEFORESTED – CONTRASTING LAND USE, IGUAÇU RIVER, BRAZIL

The contrast between native tropical rainforest and agricultural land developed after deforestation is clear in this image of the Iguaçu river in Brazil. Across the world, huge areas of forest are being cleared every year for logs, livestock or crops. World demand for soya beans in particular has led to much of this clearance. Forest, and its rich and unique diversity of wildlife, can remain untouched if it is protected for conservation within

national parks and reserves, or if it is not yet economical to clear it. In this case, the near bank of the river is part of Iguaçu National Park and is protected from deforestation.

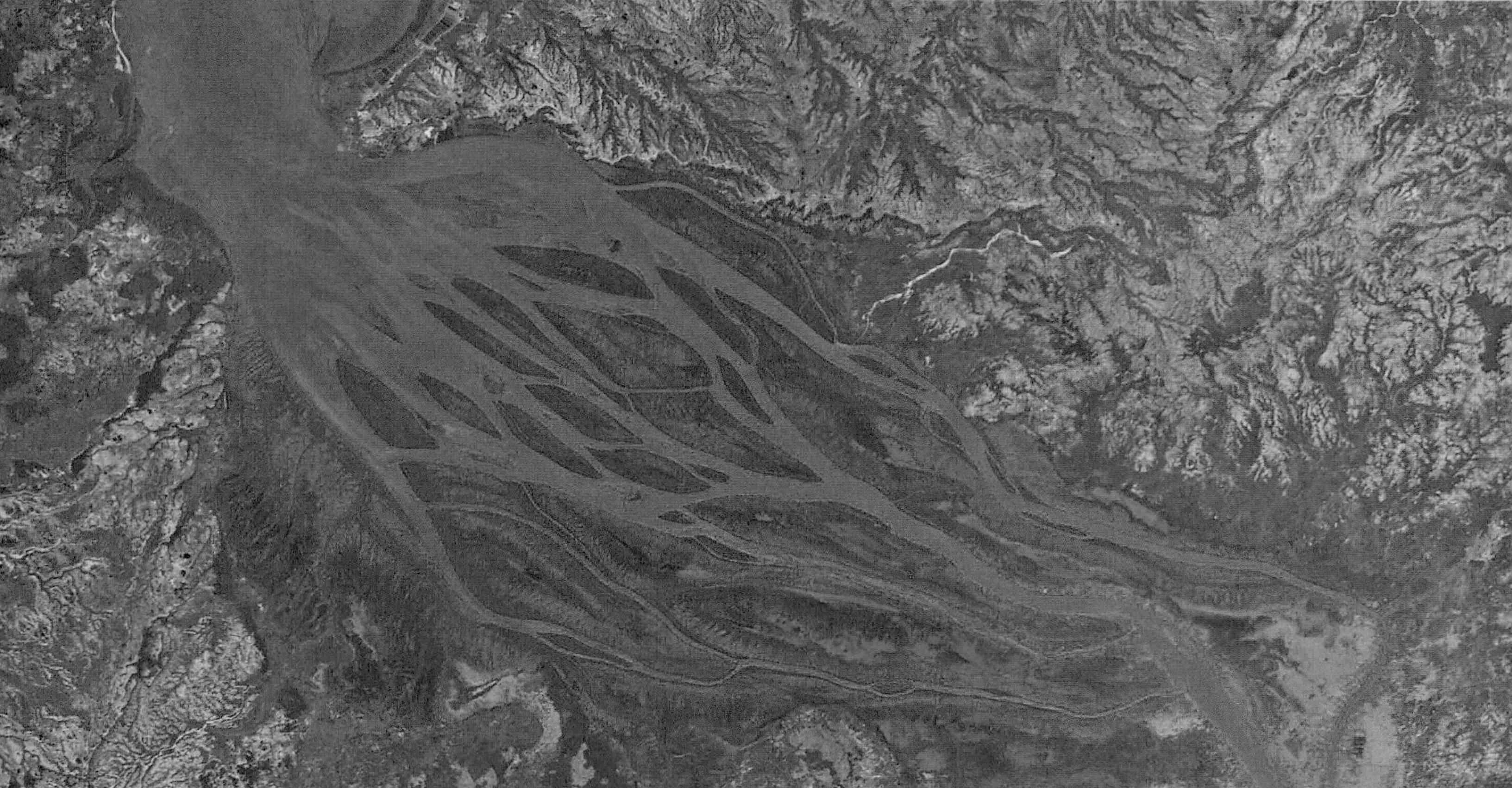

DEFORESTATION IN MADAGASCAR 2011 AND 25 MARCH 2004

Madagascar is a biodiversity hotspot with over 80 per cent of its wildlife being unique to the island. Unfortunately, rapid rates of deforestation are lowering this biodiversity; only 10 per cent of the original native forests remain. Catastrophic erosion in northwestern Madagascar has resulted from the removal of rainforests and coastal mangroves for timber. The bottom image shows the widespread flooding and massive red sediment plume in the Betsiboka river following tropical cyclone Gafilo, which hit Madagascar in March 2004. The soil washed away upstream causes the estuary to silt up.

PALM OIL PLANTATIONS, SUMATRA AND BORNEO, INDONESIA 2010

Global demand for palm oil is resulting in large areas of tropical rainforest and peat bogs on Borneo and Sumatra being destroyed to make way for vast plantations. Palm oil is used to make biodiesel and also vegetable oil used in various food and cosmetic products. By 2015 it is estimated that palm oil plantations will cover over 100 000 sq km (38 610 sq miles) of Indonesia. Loss of habitat is catastrophic to critically endangered species such as orangutans and Sumatran tigers. Indonesia is the world's third largest emitter of greenhouse gases, mainly due to deforestation.

POLLUTION –

the release into the environment of harmful or poisonous substances, often by-products of industrial or agricultural activity

The problem of solid waste – car dump, USA

POLLUTION

CO2 emissions

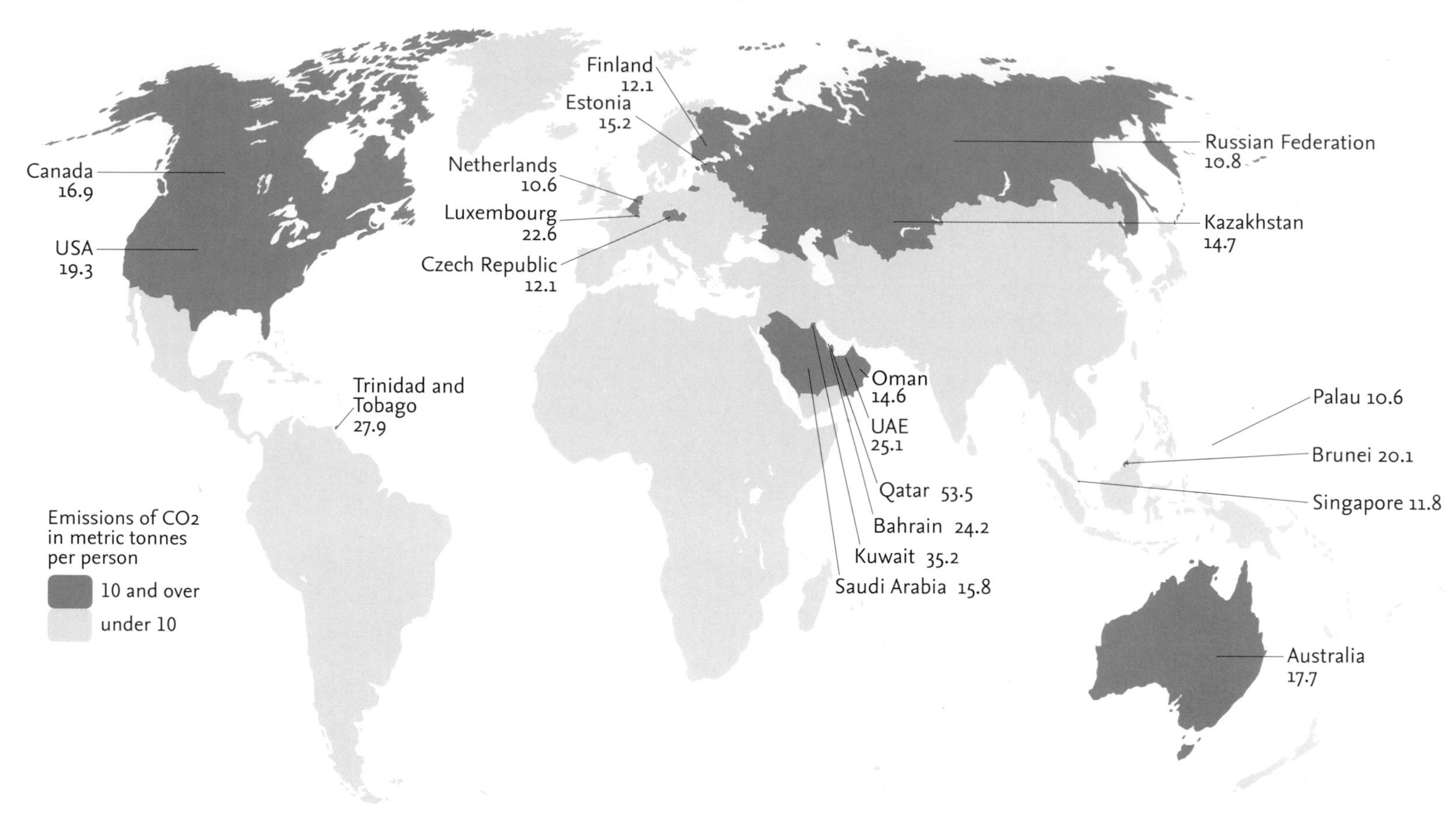

Water pollution

Emissions of organic water pollutants

Rank	Country	Grams of pollutant released per worker per day
1	**Moldova**	450
2	**Tanzania**	340
3	**Ethiopia**	240
4	**Kazakhstan**	240
5	**Tajikistan**	240
6	**Botswana**	230
7	**Cyprus**	230
8	**New Zealand**	230
9	**Mongolia**	210
10	**Eritrea**	200
11	**Kyrgyzstan**	200
12	**Ukraine**	190
13	**Azerbaijan**	180
14	**Jordan**	180
15	**Latvia**	180
16	**Macedonia**	180
17	**Bulgaria**	170
18	**Croatia**	170
19	**Lithuania**	170
20	**Mauritius**	170
21	**Russian Federation**	170
22	**South Africa**	170

Pollution, the effect of the discarded wastes of our consumer society, despoils the landscape at every scale from the small back yard to the oceans and the global atmosphere. Sometimes nature can make use of our waste, but mostly we discard it in places and in volumes which undermine natural systems and often damage human health. Air pollution, mostly from burning fuel but also from forest fires, blankets ever larger areas. Globally, 50 per cent of chronic respiratory disease is due to air pollution, notably caused by fine particles (less than 10 microns in diameter – PM10 particulates) which can penetrate deep into the respiratory tract. Millions die prematurely each year as a result. Urban smog creates acid rain which kills trees, damages crops and makes lakes and rivers toxic to fish. After initiatives to clean up the air in Europe and North America, Asia is now the most polluted continent, with winter smogs in northern India amongst the worst. A surprising amount of air pollution ends up in the Arctic. Pesticides, carried by the wind, precipitate in the cold air, sometimes reaching toxic levels in birds, whales and polar bears. The global atmosphere is also filling with carbon dioxide. Annual emissions now exceed one tonne for every inhabitant on the Earth, and atmospheric concentrations are enough to cause significant global warming.

Air quality in selected Asian cities

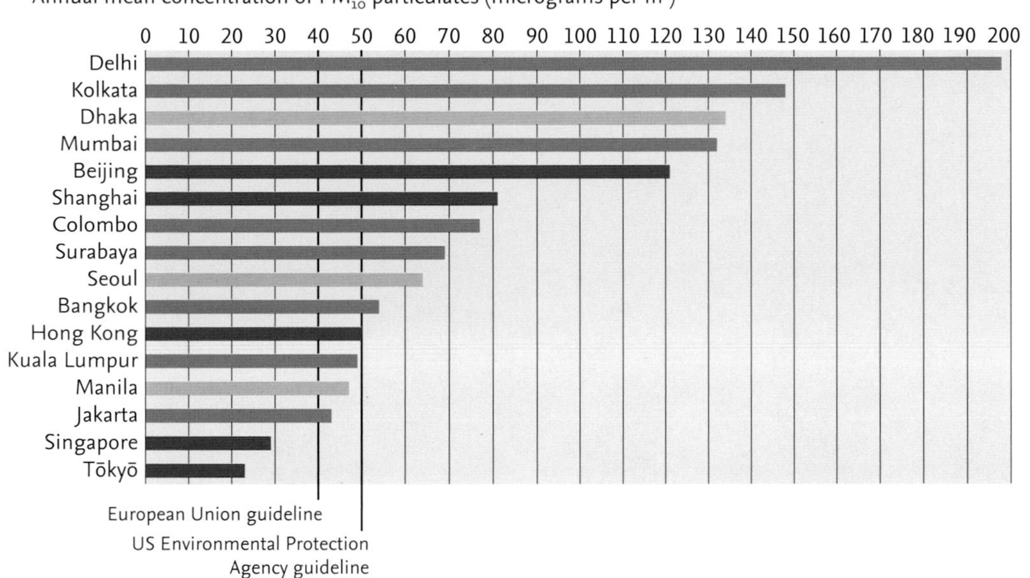

Many major rivers are heavily contaminated with sewage, industrial chemicals and fertilizer runoff from farms. Many become lifeless, and 'dead zones' are forming in the oceans at the mouths of polluted rivers such as the Mississippi. Meanwhile, soils are contaminated by landfill, factories and air pollution, and irrigated farmland in countries such as Pakistan and Uzbekistan is made infertile by salt contamination. Industrial and transportation accidents can cause major local disasters. The Exxon Valdez oil tanker spill in Alaska in 1989 did long-term damage to fragile Arctic ecosystems. The gas leak from the Bhopal chemicals plant in India in 1984 killed some 4 000 people, and fallout from the Chernobyl nuclear fire in Ukraine in 1986 may eventually be responsible for more than 10 000 deaths. The oil spill resulting from the explosion of the Deepwater Horizon oil platform in the Gulf of Mexico in 2010 did great damage to marine life and adjacent coastal wetlands.

Major pollution incidents 1967–2011

1. 1967 *Torrey Canyon* oil tanker spill, off SW England, UK
2. 1972 *Sea Star* oil tanker spill, Gulf of Oman
3. 1978 *Amoco Cadiz* oil tanker spill, Brittany, France
4. 1979 Radiation leak, Three Mile Island nuclear power station, Pennsylvania, USA
5. 1979 Ixtoc oil well blowout, Gulf of Mexico
6. 1979 Collision of *Atlantic Empress* and *Aegean Captain* oil tankers, off Trinidad and Tobago
7. 1983 Nowruz oil field blowout, The Gulf
8. 1984 Leak of toxic gas, Bhopal, India
9. 1986 Explosion and radiation leak at nuclear power station, Chernobyl, Ukraine
10. 1989 *Exxon Valdez* oil tanker spill, Alaska, USA
11. 1991 Burning oil fields during Gulf War, Kuwait
12. 1993 *Braer* oil tanker spill, Shetland Islands, UK
13. 1994 Ruptured oil pipeline, Usinsk, Russian Federation
14. 2000 Cyanide spill, Baia Mare, Romania
15. 2002 *Prestige* oil tanker spill, off NW Spain
16. 2004 Major forest fires, Spain and Portugal
17. 2005 Benzene spill, Jilin, China
18. 2010 Oil rig blowout, Gulf of Mexico
19. 2011 Following earthquake and tsunami, radiation leak at Fukushima nuclear power station, Japan

SHIPWRECK, TAURANGA, NEW ZEALAND 13 OCTOBER 2011 AND 10 JANUARY 2012

On 5 October 2011, New Zealand suffered its worst maritime environmental disaster to date when a cargo ship, the MV Rena, ran aground on the Astrolabe Reef near Tauranga. Approximately 350 tonnes of fuel-oil leaked out and washed up along the Bay of Plenty coastline, polluting the rich fishing waters and killing an estimated 20 000 seabirds. Soldiers were drafted in to join local volunteers with the beach clean up operation. Fortunately, further damage was averted as a fuel salvage operation was able to drain most of the remaining fuel on the vessel before it split in half on 8 January 2012.

NATIONAL STADIUM, BEIJING, CHINA 27 JULY 2008 AND 2 AUGUST 2008

Air pollution can be a major problem in large cities as it causes poor visibility and presents serious health risks. The pair of images above show the National Stadium in Beijing, China on a bad air pollution day and on a clear day. The stadium was the centerpiece for the 2008 Olympic Games, and in a massive effort to reduce air pollution across Beijing during the Games, the Chinese government ordered that a third of the city's cars (over 1 million), be taken off the roads, and over 100 of the most heavily polluting building sites and factories be temporarily shut down.

DEEPWATER HORIZON OIL SPILL, GULF OF MEXICO 21 APRIL AND 17 MAY 2010

Deepwater Horizon was a floating, semi-submersible drilling unit; it could operate in water depths of up to 2.4 km (1.5 miles) and drill to depths of 9.1 km (5.7 miles). On 20 April 2010 the rig exploded, killing eleven men and injuring seventeen others. The rig burned for thirty-six hours before sinking, and its damaged well began to leak tens of thousands of barrels of oil daily into the Gulf of Mexico. The natural-colour satellite image taken nearly a month after the explosion, shows a large patch of oil near the site of the accident, and a ribbon of oil stretching far to the southeast.

IMPACT OF THE DEEPWATER HORIZON OIL SPILL, GULF OF MEXICO JUNE 2010

The Gulf of Mexico oil spill was the biggest accidental marine oil spill in world history. Oil initially gushed out at a rate of about 62 000 barrels per day. The damaged well was finally capped three months later on 15 July. By then approximately 4.9 million barrels of oil had been released into the ocean causing extensive damage to the Gulf's fishing and tourism industries, and its marine and wildlife habitats. The coastline stretching from Florida to Louisiana was affected by the oil, and many animals and sea birds, such as brown pelicans, shown above, also suffered.

ATHABASCA OIL SANDS, ALBERTA, CANADA 23 JULY 1984

Athabasca Oil Sands, located in northeastern Alberta, Canada, are the world's largest oil sand deposit. Oil sands consist of clay and sand covered in water and a thick, viscous oil called bitumen. This can be processed into crude oil. In 1967 oil sand mines opened in Alberta, and with the rise in oil prices over the last decade, they have expanded dramatically. Nearly 20 per cent of the oil sands are within 75 m (246 feet) of the land's surface, and are easily extracted using open-pit mines; the remaining 80 per cent are buried deeper and are extracted using *in situ* wells.

DRAMATIC EXPANSION IN THE NUMBER OF MINES... 15 MAY 2011

By 2011, open-pit mines straddled the Athabasca River. In 2010 they covered 550 sq km (212 sq miles) and produced 356.99 million barrels of crude oil, whilst *in situ* mines produced 189.41 million barrels. Both types of mining have negative environmental impacts: extensive clearing of boreal forests and peat bogs; release of greenhouse gases, and leaching of toxins into groundwater and the river. However, Canada has strict environmental laws which ensure companies have to restore land once mining is complete. The image above shows a restored tailings pond to the west of the river.

CHEMICAL SPILL IN HUNGARY 12 OCTOBER 2010

On 4 October 2010, Hungary suffered its worst chemical accident to date. At the Ajkai Timföldgyár Alumina Plant part of a retaining wall of a waste storage pond broke and 1 million cubic m (35 million cubic ft) of toxic liquid waste was released into a local stream. The toxic sludge was a byproduct of refining bauxite into aluminium oxide (known as alumina), which has a high pH level, and led to over 120 people receiving chemical burns. The high pH also damaged local river and soil ecosystems in an area of 40 sq km (15.4 sq miles). Ten people died, most likely from drowning.

CHEMICAL SPILL IN HUNGARY OCTOBER 2010 AND 30 OCTOBER 2011

The flood of toxic chemical waste, which was 1–2 m (3–7 ft) high, inundated seven towns and villages where it pushed cars from roads, damaged houses and bridges and led to the evacuation of hundreds of local residents. The town of Devercser was one of the settlements affected and the two aerial photographs show the devastated town shortly after the flood and one year later. Houses and bridges were structurally damaged during the flood and as a result many were demolished during the clean up effort.

CORAL BLEACHING IN THE MALDIVES AND PAPUA NEW GUINEA

The images above show bleached corals in the Maldives, and partially-bleached coral in Papua New Guinea. Zooxanthellae – tiny, unicellular algae which live within coral tissues – photosynthesize and provide nutrients, and the corals' colours. When corals become stressed through changes in temperature, light or nutrients, zooxanthellae are expelled and the corals lose their colours. If the stress continues and they are not repopulated by zooxanthellae, then the corals eventually die. Over the past twenty years, coral bleaching events have increased in extent and frequency worldwide.

BYCATCH IN FISHING

Bycatch is the term given to non-target species caught in fishing equipment. Every year approximately 300 000 cetaceans (small whales, dolphins and porpoises) are killed when they are caught in fishing nets; cetaceans are mammals and can drown. Discarded fishing nets which have been cut loose from fishing vessels can pose a risk to animals since fish, mammals and turtles can still become trapped in the drifting nets and will drown or die of starvation. In the photographs above a Hawaiian Monk Seal and a turtle have become trapped in discarded netting.

ON LIFE SUPPORT

John Grace

About 150 hectares (370 acres) of rain forest go up in smoke each hour. On those naked hectares farmers will sow tropical grasses to raise beef, or they might plant soya bean, or sugar cane or perhaps palm oil. Mostly these products will be sold on the world markets for people like you and me.

Each hour this activity will release greenhouse gases, including around 100 000 tonnes of carbon dioxide. On top of that, forests absorb more solar energy than crops, and they humidify and cool the atmosphere. When they are replaced by crops, much of that benefit is lost. Their tree-tops are homes for perhaps half the world's animal species, some of them still unknown to science. Ultimately, these species will be lost too. If we consider all the benefits of forest, as a regulator of climate, a provider of materials, places to live, and a gene bank, we conclude that forests are a valuable resource, and that trees are more valuable alive than dead.

The real deforestation rate is probably greater than the official figures suggest, as forests are becoming degraded by logging. This process involves loss of large trees, which are often key habitats for animal species. Only in recent years has satellite remote sensing been able to distinguish between degraded and intact forest, and even now, much of the official data on deforestation is based on out of date information submitted by governments in countries where satellite observations are not made. In the future, it is expected that radar satellites will be employed: they can see through clouds and as the radar penetrates into the canopy they can be used to estimate the extent to which the forest is degraded. Already we are beginning to see that the official statistics under-estimate the extent to which the Earth's green mantle is being eroded.

An important issue for the future is how the remaining rain forest can be conserved for future generations. Before the invention of agriculture, there may have been only a few million human inhabitants on the whole planet, fewer than those living today in London, New York or São Paulo. Then, much of the land was densely forested. Now, there are over seven billion people, much more land is needed to grow food, and the forests are fast becoming mere fragments. The human population may reach ten billion by 2050. No one really knows the ultimate carrying capacity of the planet. We cannot estimate how the need for so much food will impinge on forests beyond that date, especially when one considers the necessary geographical shifts in the world's farmlands as warming occurs, the political pressures that this will entail, and the trend of increasing the per capita meat consumption in China and India.

'Pollution' is a blanket term that includes contamination of water and the atmosphere by the various products of human activity. In earlier times pollution was seen as a local phenomenon. Rachel Carson's book *Silent Spring*, published in 1962, made the point that pollution, even then, was a global phenomenon. Today, the focus is on greenhouse gases. Fossil fuel burning and deforestation continue apace, and thus the concentrations of carbon dioxide in the atmosphere are rising, contributing to dangerous global warming and acidification of the ocean. There has been some progress. Some of the pernicious pesticides have been banned and the production of chlorofluorocarbons has been curtailed. Industrial processes have become cleaner and cars are substantially more efficient. Much of these improvements come from new research and the brave actions of the environmental movement. However, the task is not complete. We still have major pollution by nitrogen oxides, heavy metals, photochemical smog, radioactivity and diverse industrial emissions that interfere with biological functions and damage ecosystems. To some extent, the rich countries have exported their industrial activity and pollution emissions to China, India and the other 'tiger economies' of the East. Low-paid workers in those countries make the white goods and electronic components on behalf of all of us. Tackling deforestation and pollution costs money and needs to be a priority for governments. The challenge is enormous. The UN-REDD Programme (The United Nations Collaborative Programme on Reducing Emissions from Deforestation and Forest Degradation in Developing Countries) is an attempt to organize payments to tropical countries to halt deforestation, but it isn't yet working. The Kyoto Protocol was initiated to reduce greenhouse gas emissions, but these emissions continue to rise inexorably. Both initiatives do, however, mark the beginning of a process whereby the richer countries take joint responsibility for the life support system of Planet Earth.

Northeren Italy is one of Europe's more polluted areas. This image shows smog hovering in the area south of the Alps.

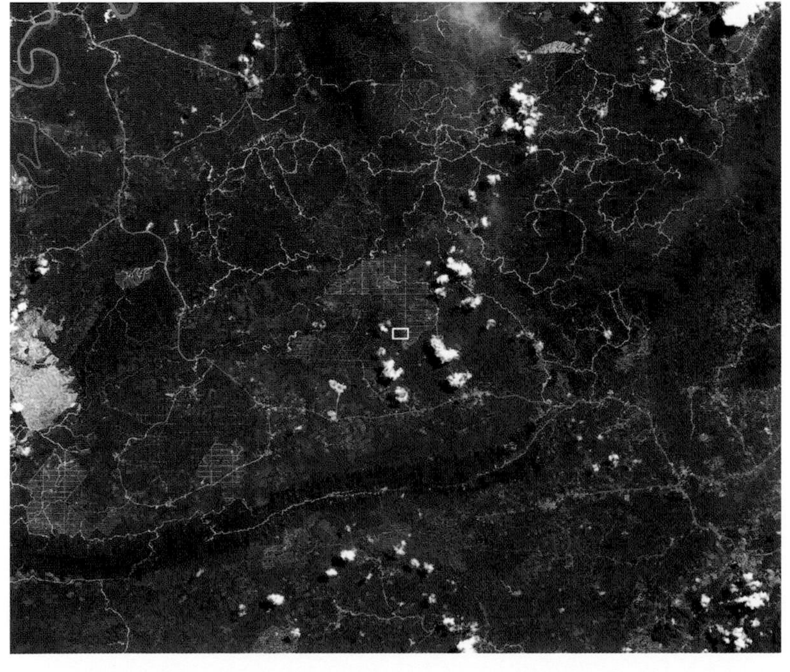

The high resolution natural colour satellite image above shows land use changes on the edge of Borneo's rainforest – remaining areas of dense native forest, cleared land, more sparsely covered palm oil plantation and access roads.

The satellite image to the left shows the location of the top image at the edge of a plantation. Intact forest shows as dark green with roads and rivers in light brown.

WARMING

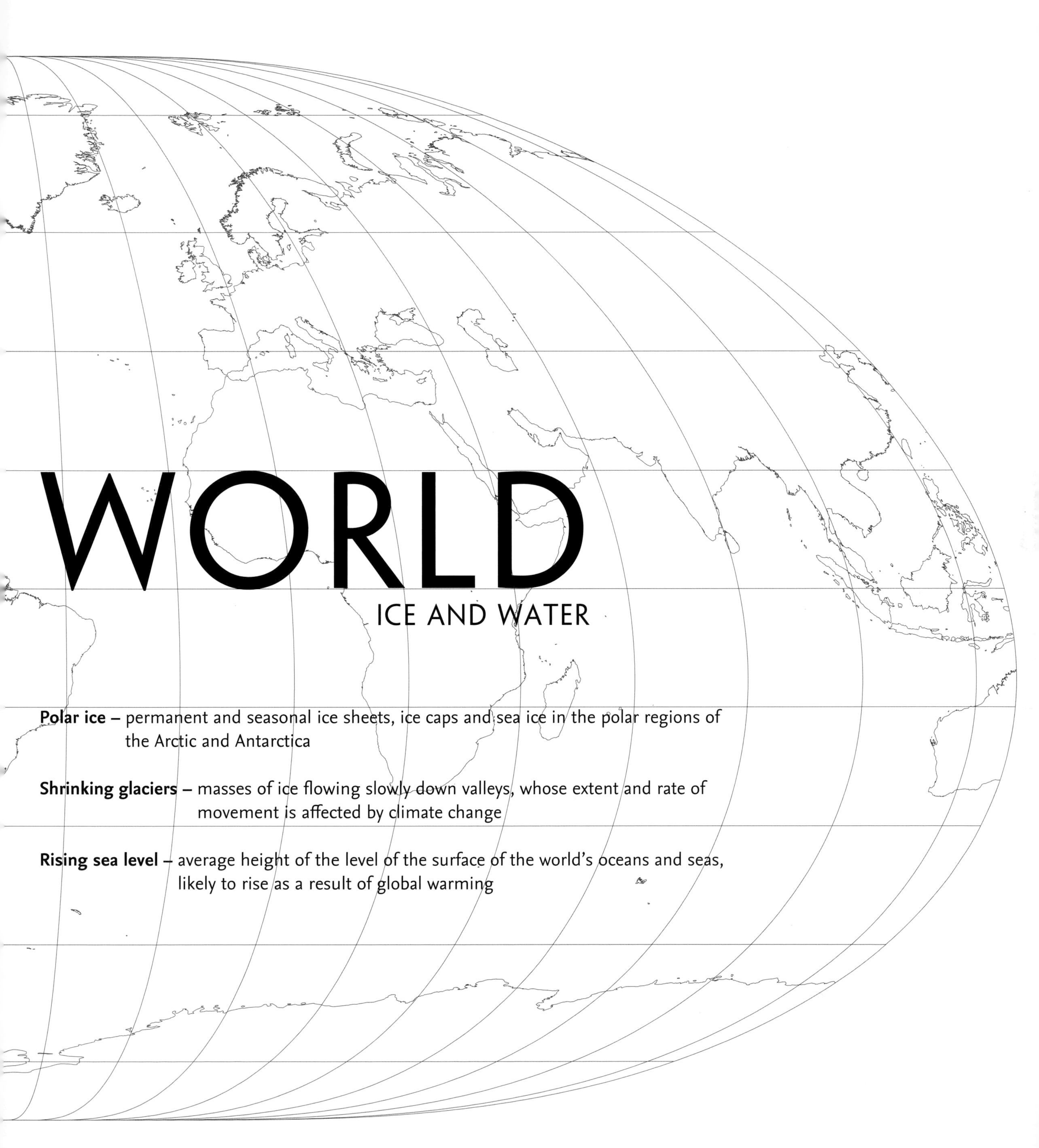

WORLD

ICE AND WATER

Polar ice – permanent and seasonal ice sheets, ice caps and sea ice in the polar regions of the Arctic and Antarctica

Shrinking glaciers – masses of ice flowing slowly down valleys, whose extent and rate of movement is affected by climate change

Rising sea level – average height of the level of the surface of the world's oceans and seas, likely to rise as a result of global warming

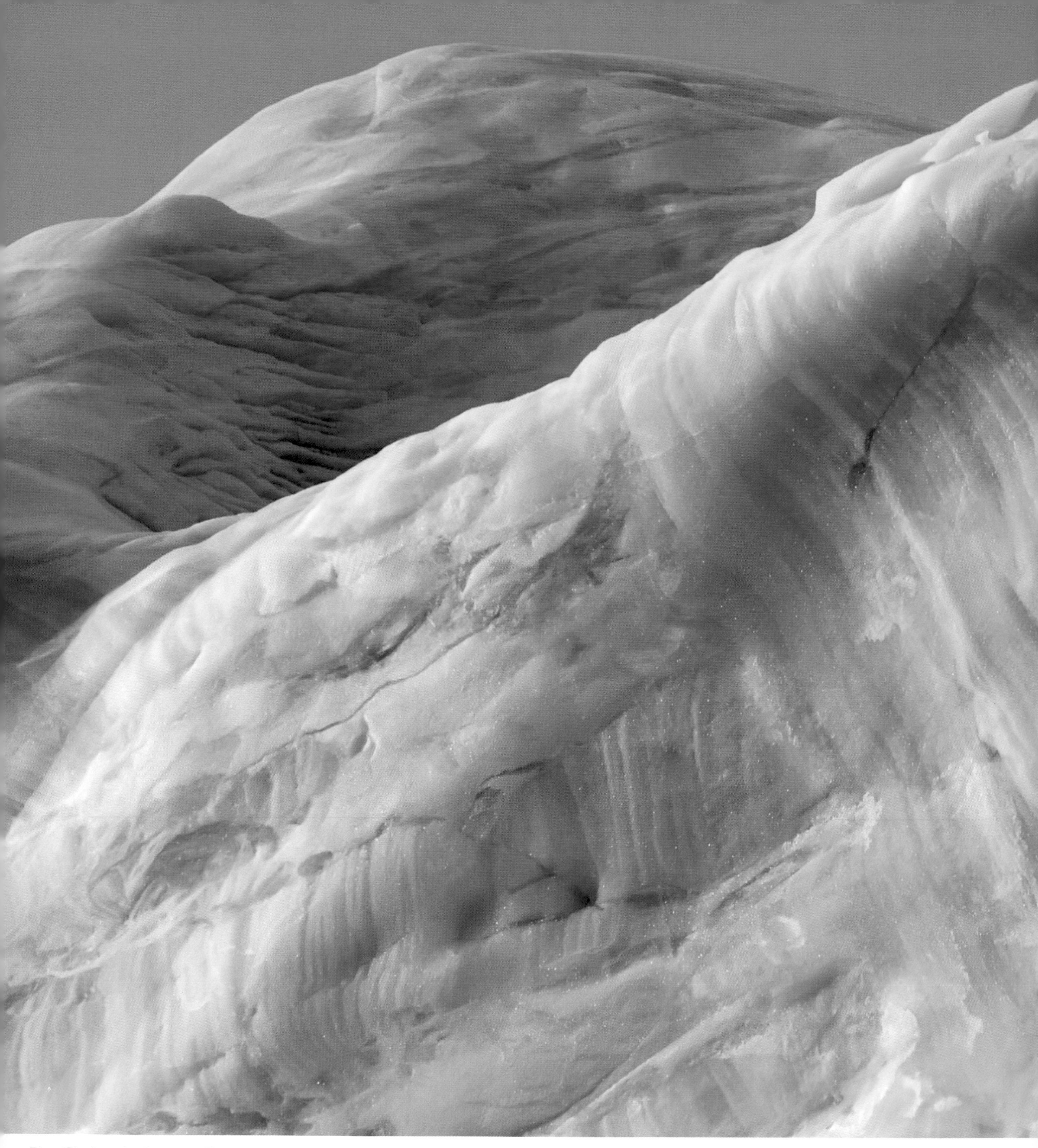

POLAR ICE –

permanent and seasonal ice sheets, ice caps and sea ice in the polar regions of the Arctic and Antarctica

Blue Iceberg

POLAR ICE

The northern and southern polar ice caps are very different. Although the Arctic includes land areas such as Alaska, Greenland and Siberia, the North Pole itself sits in the middle of the Arctic Ocean, covered only by drifting sea ice a few metres thick. At the South Pole, near the centre of the Antarctic continent, the presence of a huge landmass has allowed a colossal ice sheet, up to 4 km (2.5 miles) thick, to build up.

The Antarctic supports barely any plant life, whereas the Arctic is host to massive boreal forests and – further north still – a wide extent of treeless tundra. The Arctic is also home to nearly four million people, whereas the southern continent is host only to scientific research stations and a few wandering explorers. Both support abundant animal life, with polar bears in the Arctic and penguins in the Antarctic being the most iconic species.

Antarctica

Area	sq km	sq miles
Total land area *(excluding ice shelves)*	12 093 000	4 667 898
Ice shelves	1 541 700	595 253
Exposed rock	49 000	18 914
Heights	**m**	**ft**
Lowest bedrock elevation *(Bently Subglacial Trench)*	-2 496	-8 189
Maximum ice thickness *(Astrolabe Subglacial Basin)*	4 776	15 669
Mean ice thickness *(including ice shelves)*	1 859	6 099
Volume	**cubic km**	**cubic miles**
Ice sheet including ice shelves	25 400 000	6 094 000
Climate	**°C**	**°F**
Lowest screen temperature * *(Vostok Station, 21 July 1983)*	-89.2	-128.6
Coldest Place – annual mean *(Plateau Station)*	-56.6	-69.9
Mean annual temperature at South Pole *(Amundsen-Scott Base)*	-49.5	-57.1
Lowest recorded temperature at South Pole *(Amundsen-Scott Base)*	-66.8	-88.2
Highest recorded temperature at South Pole *(Amundsen-Scott Base, 25 Dec 2011)*	-12.3	9.9

* lowest temperature ever recorded on Earth

Global surface temperatures 1880–2010

Four independent records show nearly identical long-term warming trends

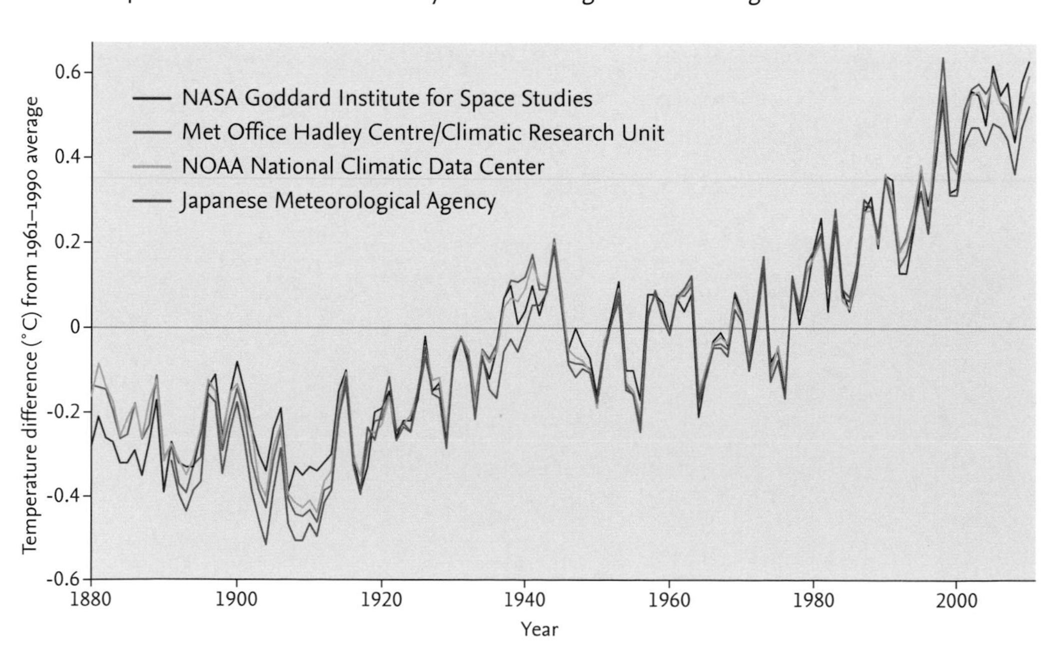

Global warming is exaggerated at the poles, and in the Arctic temperatures have risen at twice the global average rate over the last few decades. The main reason is the 'ice-albedo' effect, where melting snow and ice reveal darker ocean and land surfaces underneath, absorbing more of the sun's energy and causing more warming in a self-reinforcing cycle.

The Antarctic is insulated from wider global changes because it is surrounded by circumpolar ocean currents and winds, maintaining very low temperatures in the interior. Although climate records are sparse, Antarctica's surface is not thought to have warmed significantly over the last thirty years. The exception is the Antarctic Peninsula, which extends north towards South America and is therefore more exposed to the changing climate. The western side of the Peninsula has records going back over fifty years, which show an extraordinary 3C° (5.4F°) rise in average temperatures.

The Arctic

Precipitation	mm	inches
Average precipitation *(mainly snow)* **in the Arctic Basin – rain equivalent**	130	5.1
Average precipitation *(mainly snow)* **in the Arctic coastal areas – rain equivalent**	260	10.2
Ice thickness	**m**	**ft**
Average sea ice thickness	2	6.6
Maximum ice thickness *(Greenland ice sheet)*	3 400	11 155
Sea ice extent	**sq km**	**sq miles**
Minimum sea ice extent in summer	5 000 000	1 930 000
Maximum sea ice extent in winter	16 000 000	6 180 000
Temperature	**°C**	**°F**
Summer temperature at North Pole	near 0	32
Winter temperature at North Pole	-30	-22
Lowest temperature recorded in the Arctic *(Verkhoyansk, northeastern Siberia, 1933)*	-68	-90

The changing Arctic

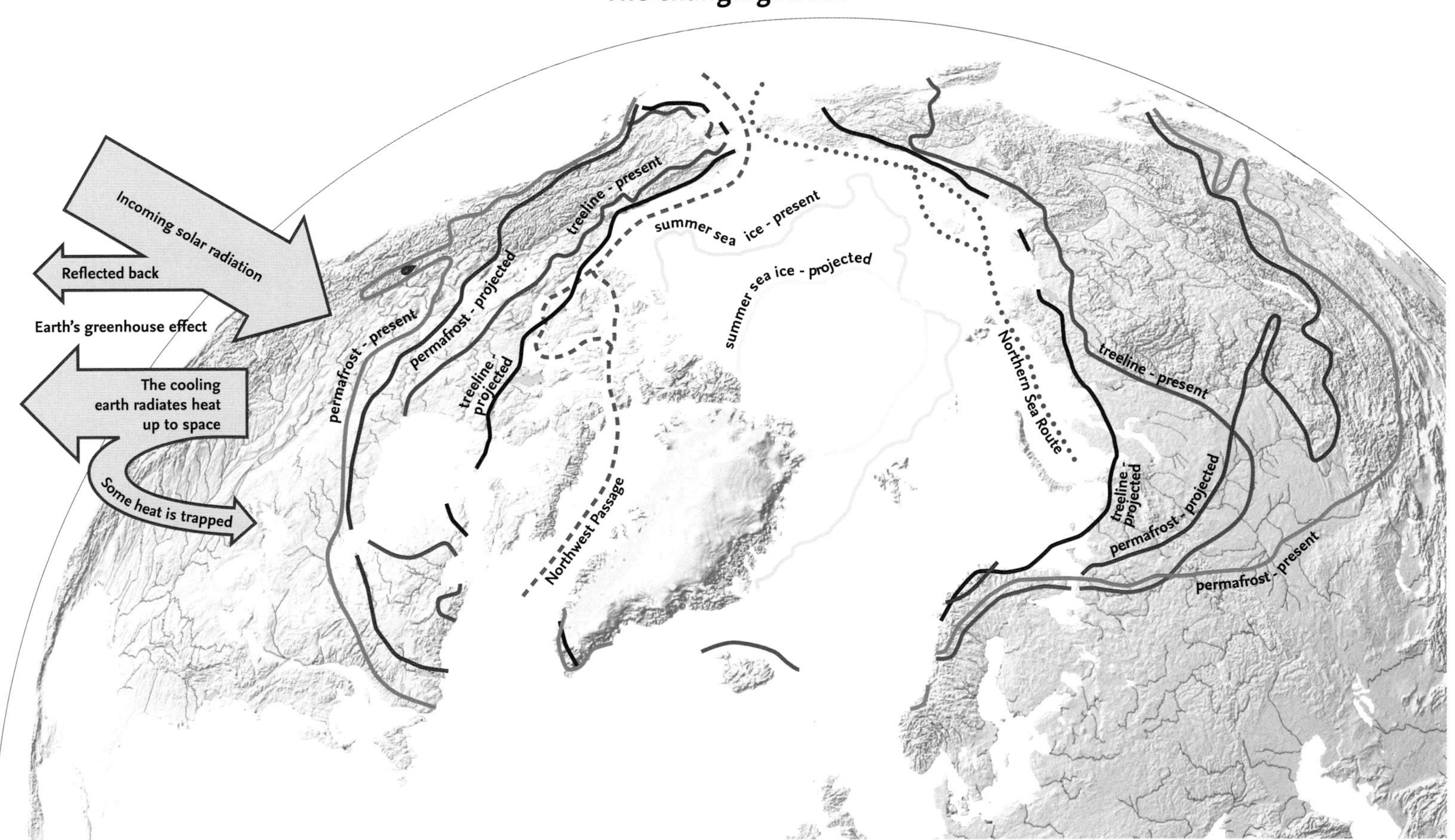

Antarctic profile

Cross-section of West Antarctica from the Ronne Ice Shelf to the Ross Ice Shelf

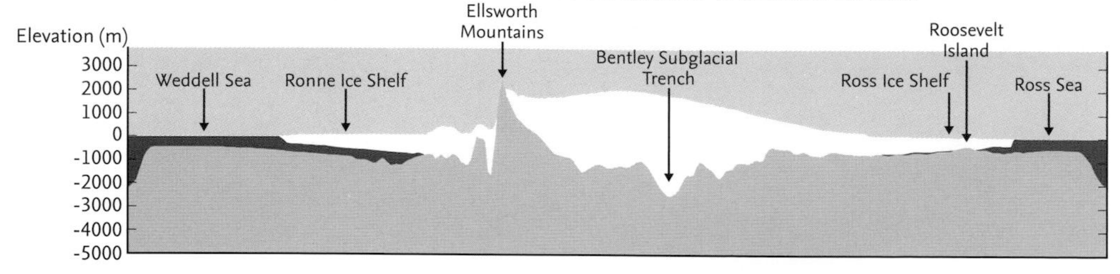

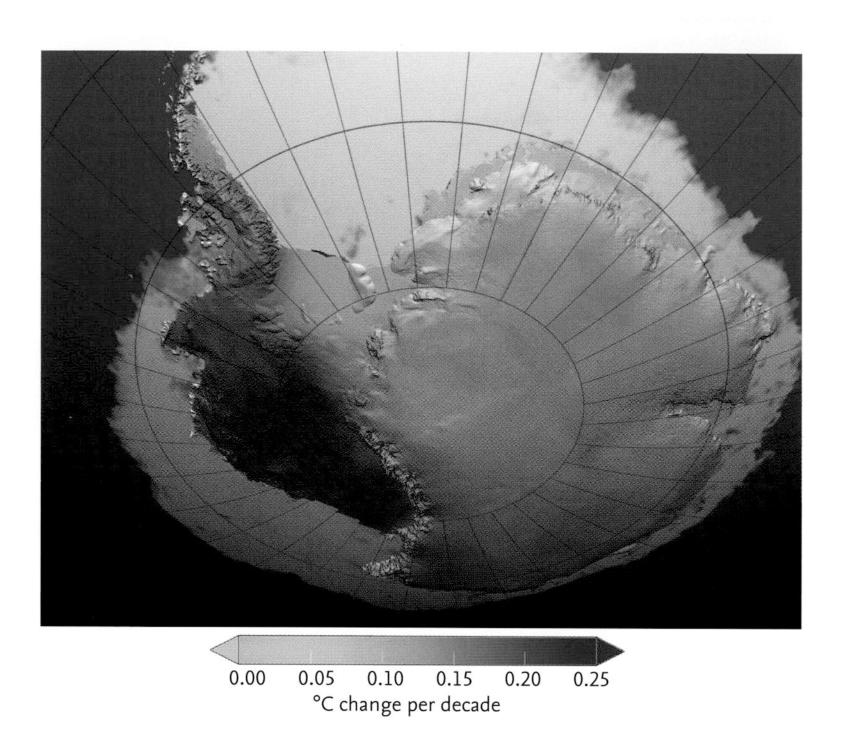

Although the greatest loss of sea ice is in the Arctic, and the bulk of the Antarctic ice sheet has not warmed significantly, there is unmistakable evidence of a warming climate in parts of Antarctica. Warmer sea temperatures have thinned the West Antarctic ice sheet between the Antarctic Peninsula and the Ross Ice Shelf, causing an increase in sea ice around the continent. The most rapid warming is inland from the Antarctic Peninsula with the last fifty years seeing an increase of over 0.1°C (0.18°F) in surface temperature every ten years.

If all of Greenland melted, it would add 7 m (23 feet) to global sea levels, although it is thought that this would take several centuries. Greenland's major glaciers are already thinning and speeding up, however, suggesting that this meltdown is beginning to happen. Moreover, the surface melt area on the top of the ice cap is also increasing, contributing to rivers and meltwater lakes during the summer months.

Arctic sea ice reaches its annual minimum extent in September, with progressively earlier breakup and lengthening melt seasons being recorded in many areas. Records indicate that Arctic sea ice extent has been declining since at least the early 1950s, with a decline of 12 per cent every ten years since 1979. September 2007 saw the lowest levels of Arctic sea ice ever recorded, at a level 23 per cent lower than the previous record in 2005. This record low was very nearly matched again in 2011 prompting predictions that the Arctic could be ice-free within thirty years. As well as a decline in extent, records show that cover has become thinner, creating a more vulnerable perennial cover. Climate models project a total loss of summer sea ice in perhaps only a few decades, spelling disaster for ice-dwelling animals such as polar bears.

Arctic Ocean profile

Cross-section of the Arctic Ocean from northwest Canada to northwest Russian Federation

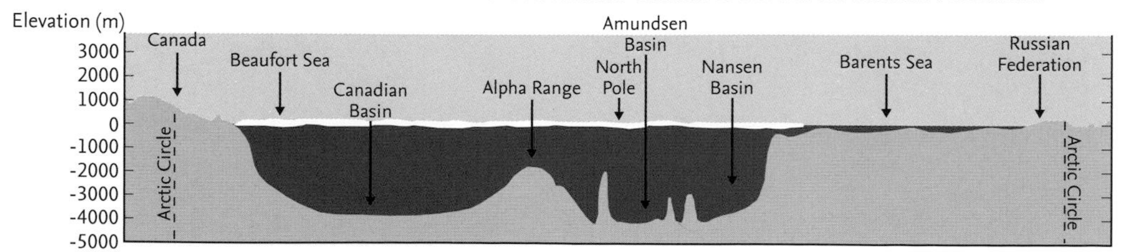

Arctic sea ice concentration September 1980

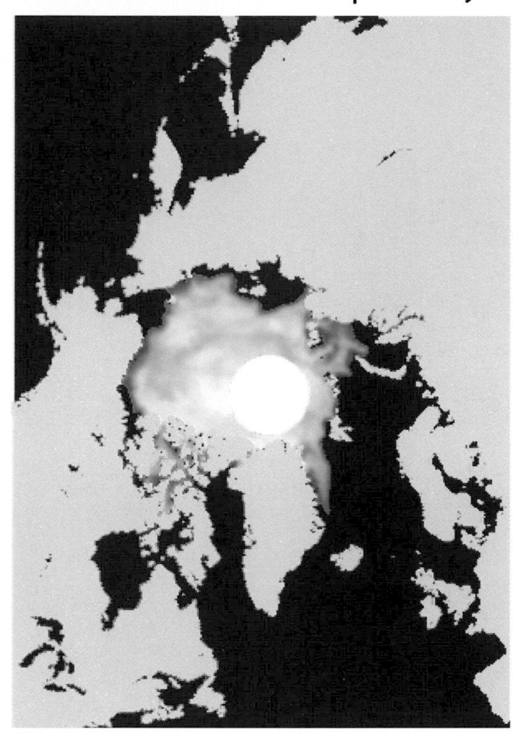

Arctic sea ice concentration September 2011

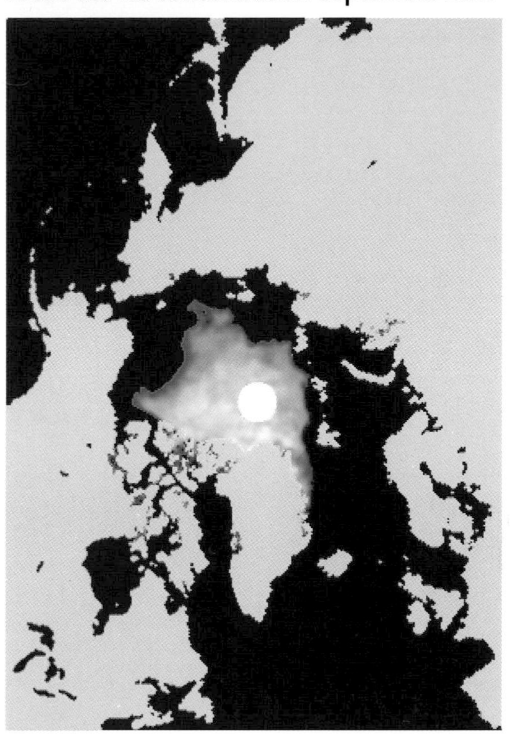

Antarctic sea ice concentration February 1980

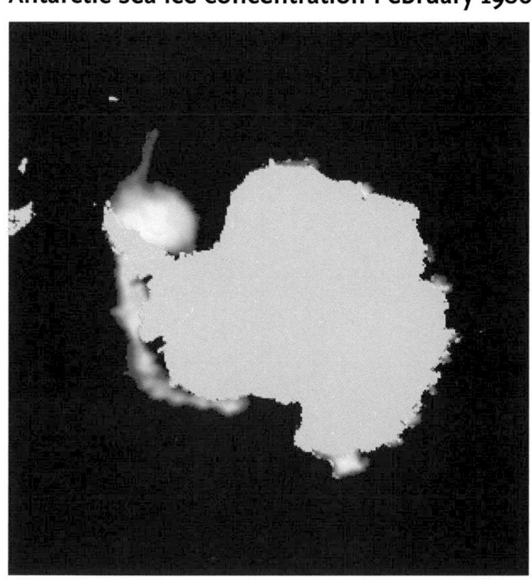

Antarctic sea ice concentration February 2012

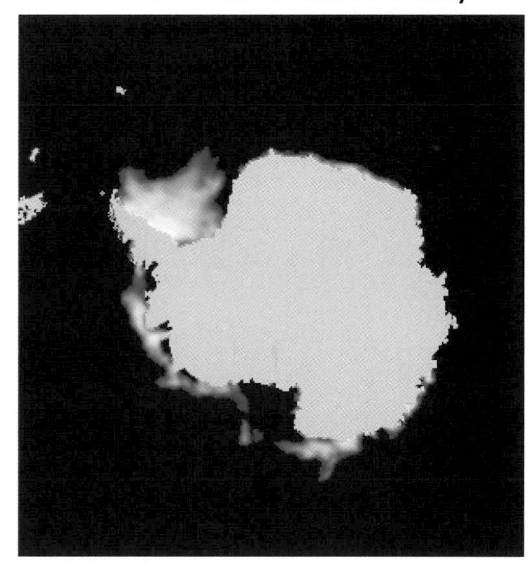

per cent

0 5 10 15 20 25 30 35 40 45 50 55 60 65 70 75 80 85 90 95 100

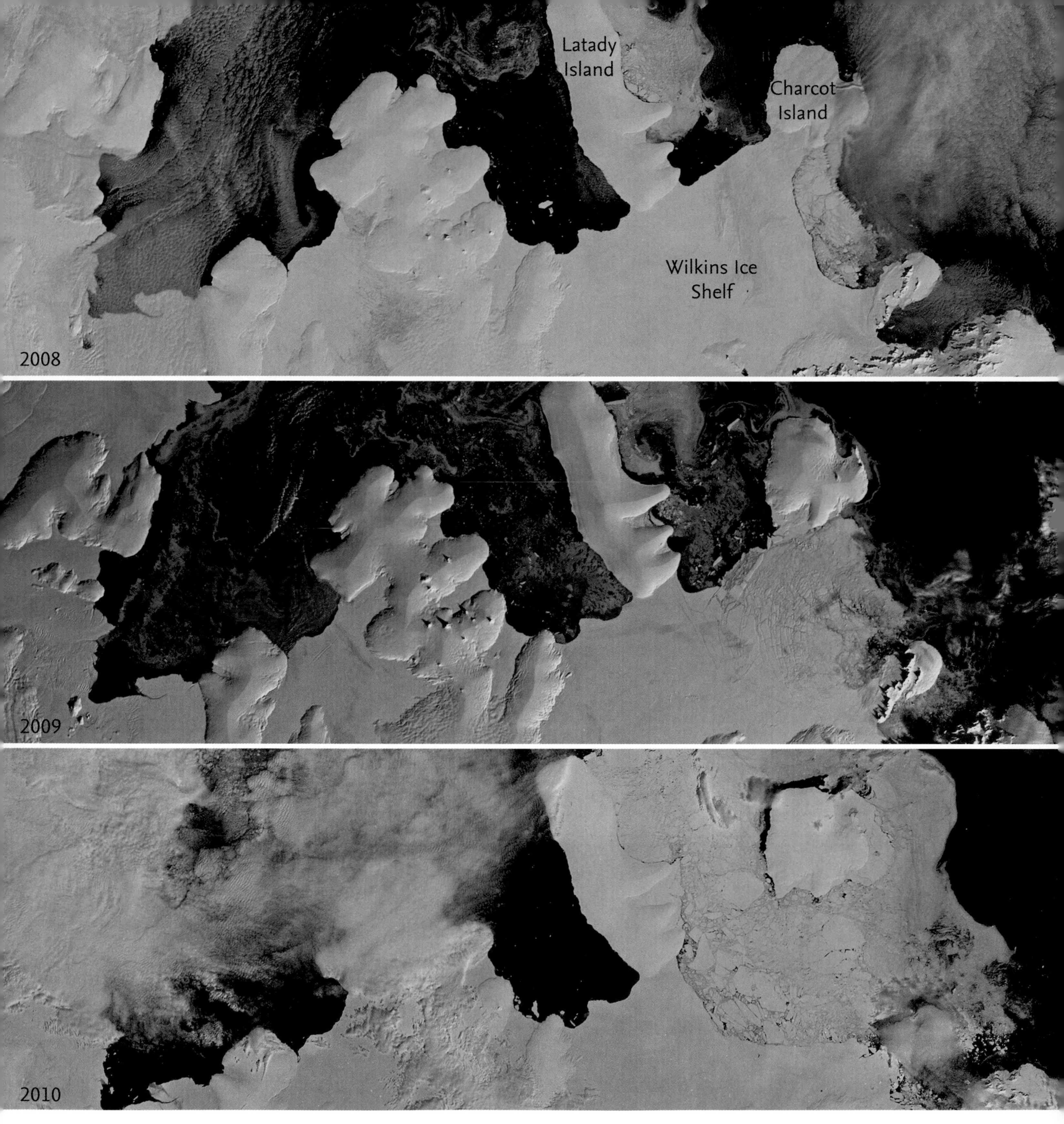

WILKINS ICE SHELF, ANTARCTIC PENINSULA 28 FEBRUARY 2008, 10 APRIL 2009 AND 7 JANUARY 2010

The Wilkins Ice Shelf used to extend from the western edge of the Antarctic Peninsula to Latady and Charcot islands. It had been stable throughout most of the twentieth century, however in the 1990s it began to retreat. In 2008 there were multiple disintegration events, and by 2009 most of the western and northern sections of the ice shelf had collapsed; only a narrow ice bridge remained intact to connect it to Charcot Island. However, in early April 2009 the ice bridge disintegrated, leaving the remaining intact southern portion of the ice shelf more vulnerable to collapse in the future.

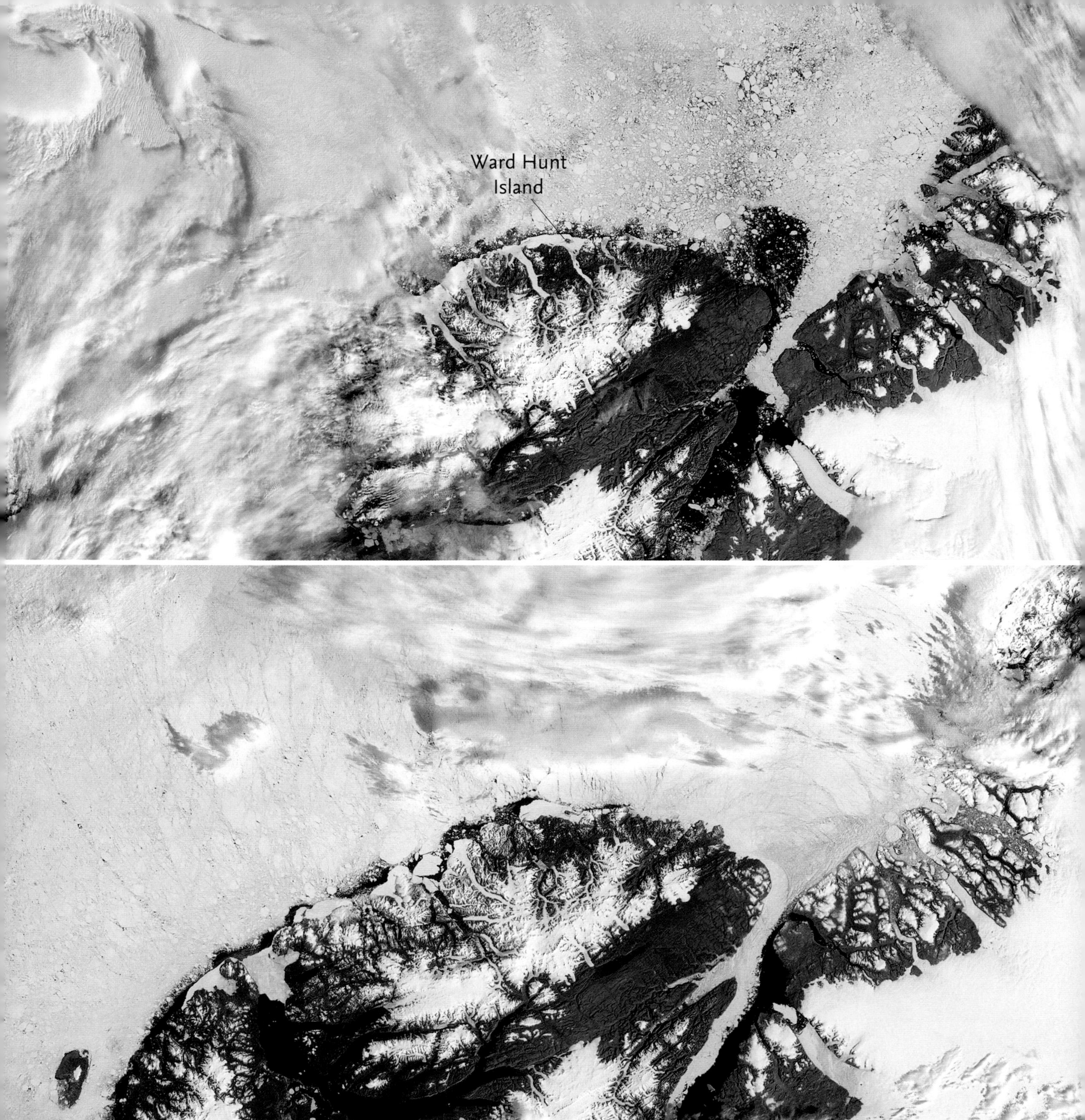

WARD HUNT ICE SHELF, ELLESMERE ISLAND, CANADA 22 AUGUST 2002 AND 18 AUGUST 2010

An ice shelf typically advances for several decades until it becomes unstable and icebergs break off, or calve, from the front of the shelf; this advance and retreat is normal and maintains the ice volume. However, stages of rapid disintegration into small pieces can indicate warming temperatures.

The Ward Hunt Ice Shelf is one of several on Ellesmere Island to retreat in the last decade. The 2010 satellite image shows ice-free fjords to the south of the ice shelf, along with fractures and disintegrating ice to the east and south of Ward Hunt Island, which is in the middle of the ice shelf.

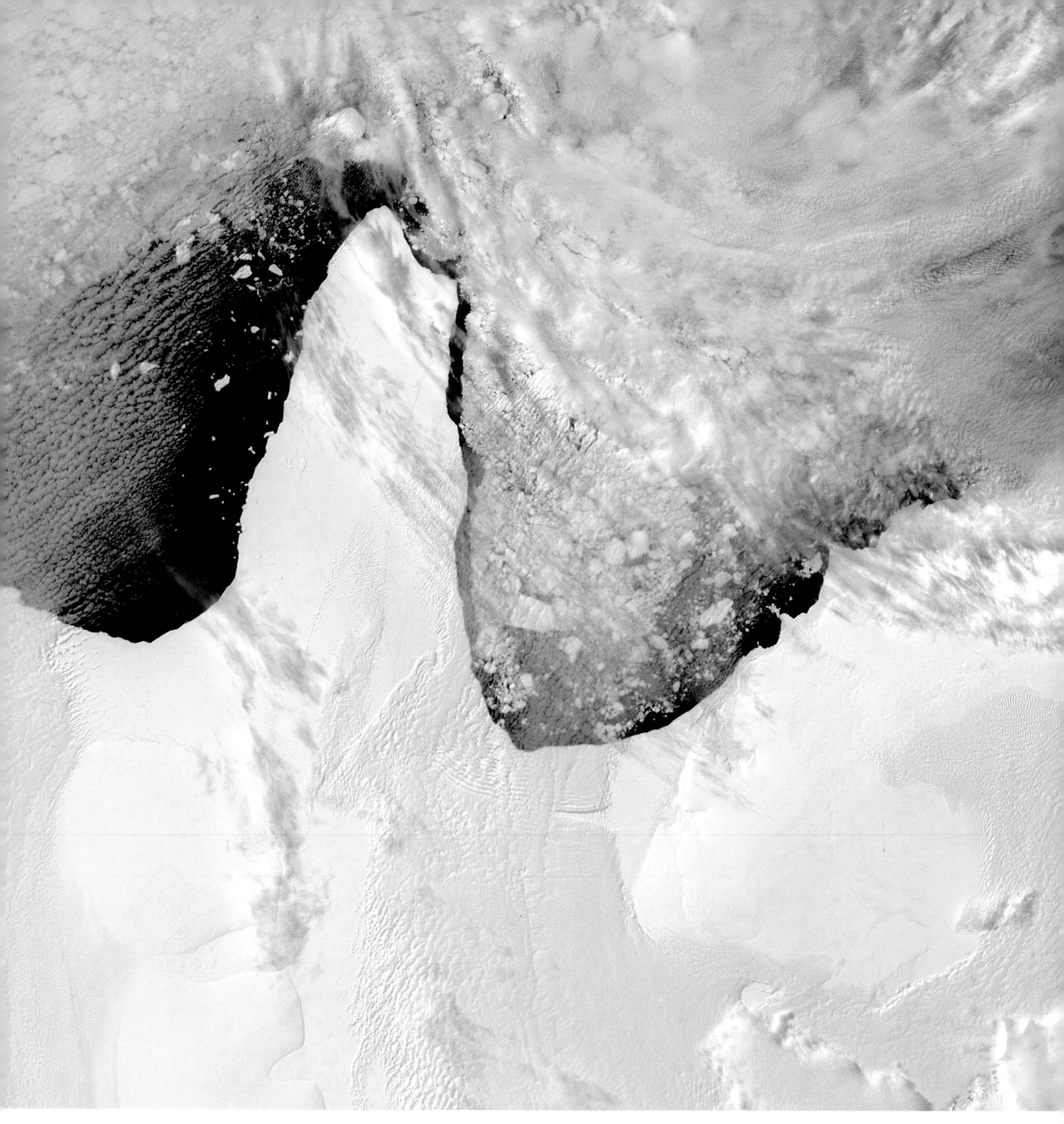

RONNE-FILCHNER ICE SHELF, ANTARCTICA 12 JANUARY 2010

This ice shelf bordering the Weddell Sea, covers an area of around 430 000 sq km (166 000 sq miles) and is the second largest in Antarctica after the Ross Ice Shelf. The shelf is divided into the Western (Ronne) and Eastern (Filchner) sections which grow due to the flow of the inland ice sheets. When the stress on the ice gets too great, cracks form and parts of the ice sheet break away from the ice shelf creating icebergs. This process is known as calving.

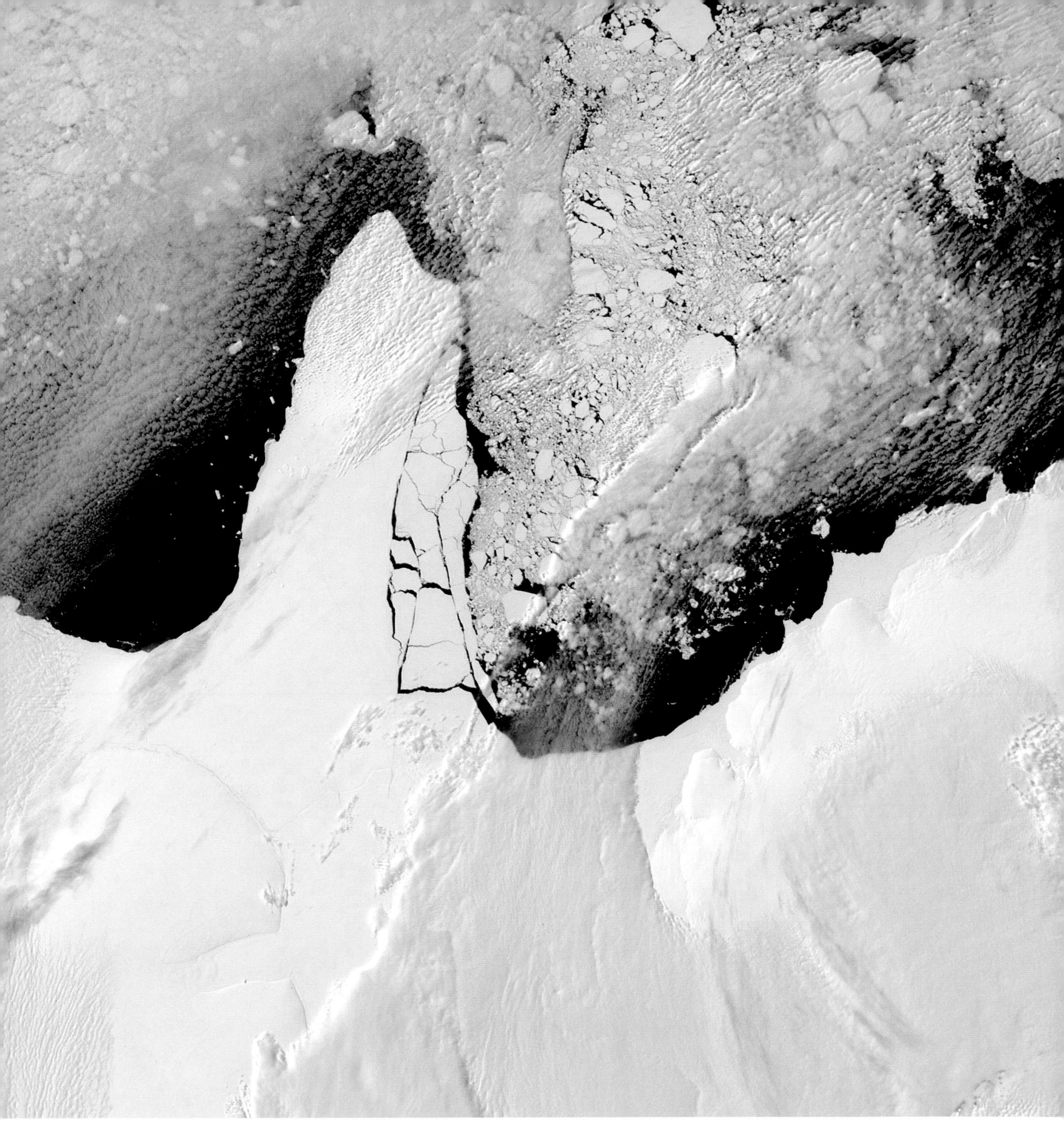

RONNE-FILCHNER ICE SHELF, ANTARCTICA 13 JANUARY 2010

In the day between these two images an area around 4000 sq km (1544 sq miles) broke away from the ice shelf and disintegrated into smaller pieces. The narrow tongue of ice where the break-up happened is a bridge of thin sea ice which is anchored to the shore. This ice bridge regularly breaks up and reforms and is a dramatic introduction to the arrival of summer in Antarctica.

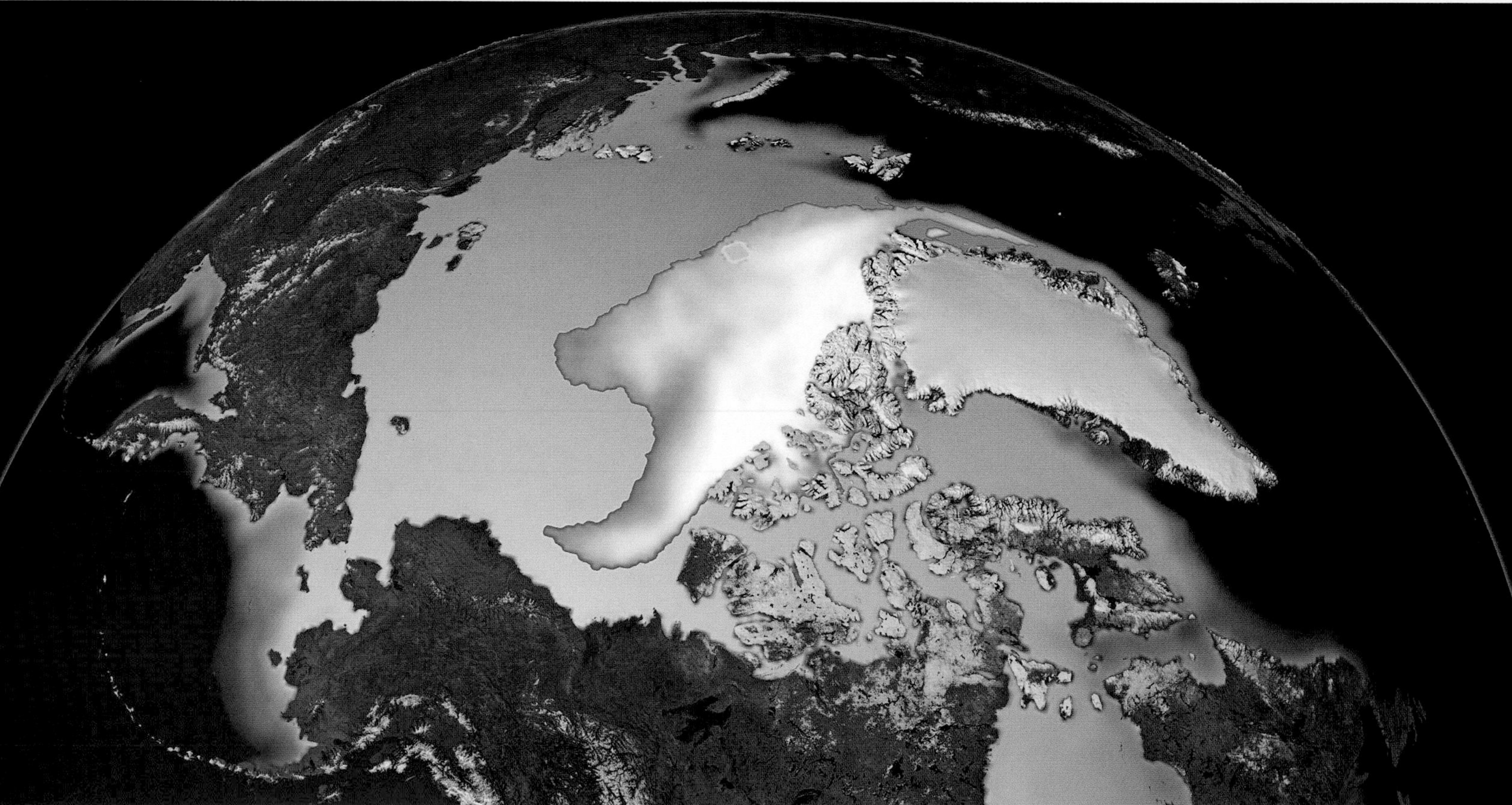

REDUCTION IN OLDEST ARCTIC ICE SHELF WINTER 1979–1980 AND WINTER 2011–2012

Passive microwave satellite data is used to monitor sea ice cover and thickness. In the last three decades it has become clear that the oldest and thickest Arctic sea ice is thinning and retreating at a quicker rate than the younger and thinner ice at the edges of the ice cap. Sea ice coverage is shown in the images for 1980 (top) and 2012 (bottom). Average sea ice cover is shown in light blue to milky white, while older, multi-year ice is shown in bright white. The Arctic multi-year sea ice extent is decreasing at a rate of 15.1 per cent per decade.

SEA ICE OFF THE NORTHERN COAST OF GREENLAND

The top image shows fractured sea ice. A combination of powerful waves, large rocks and the shallowness of the water led to the sea ice breaking into large pieces. The bottom image shows 'pancake ice', which is an early stage of sea ice development. When ocean waters become cold enough to begin freezing, small ice crystals called 'frazil' form. If the ocean is rough, 'frazil crystals' will amalgamate to form circular disks called 'pancake ice'. Eventually the 'pancakes' pile up on top of one another and combine to form a solid ice sheet.

WINTER AND SUMMER VIEWS OF SHISHMAREF, ALASKA, USA 2003 AND 2010

The Inupiaq village of Shishmaref is located on Sarichef Island in the Chukchi Sea, just north of the Bering Strait and 8 km (5 miles) from mainland Alaska. For much of the year sea ice gives protection to the shoreline from erosion by waves and storm surges. It used to be present during the summer months, but this is no longer the case, and the coastline is suffering as a result. Every year approximately 3 m (10 feet) of Shishmaref is eroded by the sea. Sea walls were built to give the village some of the protection once afforded by the sea ice.

EFFECTS OF MELTING PERMAFROST, SHISHMAREF, ALASKA, USA 2003 AND 2010

Rising temperatures are causing the permafrost to melt. This destabilizes the shoreline making it much more vulnerable to erosion. As the permafrost thaws it loses its capacity to support heavy structures, and the coastline crumbles into the sea taking houses with it. This was the case in the top photograph, which shows two residents at the site of their former home. Residents built any new houses on the island's old runway as it provided firmer foundations. In 2002 the community voted to relocate the village to the mainland, but funding issues have delayed the process.

UUNARTOQ QEQERTAQ, GREENLAND 11 AUGUST 1985 AND 4 SEPTEMBER 2005

In 2005 a new island, Uunartoq Qeqertaq or Warming Island, became visible off the northern end of Liverpool Land, a peninsula on Greenland's eastern coast. The island was previously covered in a thick layer of ice, and scientists assumed that it was just part of the peninsula. Rising temperatures across the Arctic are responsible for ice melting around Greenland's coastline. In the case of Uunartoq Qeqertaq, approximately 10 km (6 miles) of ice retreated over a five year period to reveal the island in 2005.

WILDLIFE IN THE ARCTIC AND ANTARCTIC

The effect of melting Arctic ice on polar bears is significant. They are forced to spend longer on land, the ice flows they rely on to move around are reduced and they can drown or become stranded and starve. Across the Antarctic Peninsula temperatures have increased by 3°C (5.4°F) over the past fifty years. This has led to a marked decrease in sea ice, which in turn has led to a reduction in numbers of krill as it is vital to their life cycle. Krill are the Chinstrap penguins' main food source, and several colonies have seen a reduction in numbers of 30–66 per cent due to food shortages.

SHRINKING GLACIERS –

masses of ice flowing slowly down valleys, whose extent and rate of movement is affected by climate change

Cerro Torre, Patagonia, Argentina

SHRINKING GLACIERS

Global distribution of glaciers and ice sheets

Glaciers numbered are those shown in the graph on the opposite page

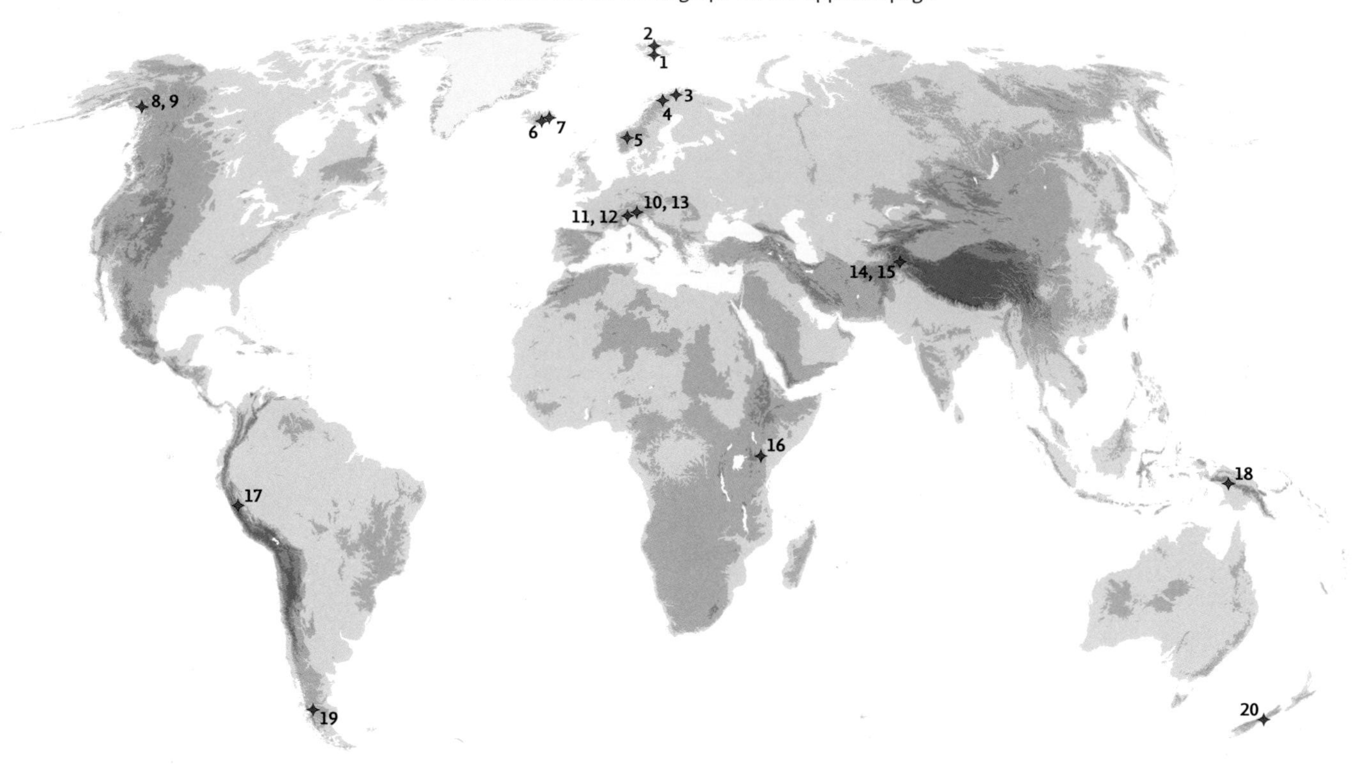

Glaciers form wherever it is cold enough for snowfall to accumulate into thick layers of ice over many years rather than melting away during the summer. Areas cold enough to support permanent ice may be close to sea level at high latitude regions such as Patagonia, but only exist at high altitudes of 5000 m (16 404 feet) or more in hot tropical areas such as Peru.

Glaciers expand and contract according to the balance between temperature and precipitation. Like a bank account, withdrawals from melting must be balanced by deposits from new snowfall or the glacier will shrink. Glaciers also shrink if the melt area on their surface increases because rising temperatures push the snowline higher up the mountainside. This is the main way in which global warming is affecting glaciers.

Cumulative changes in global glacier thickness and glacier contributions to sea level rise

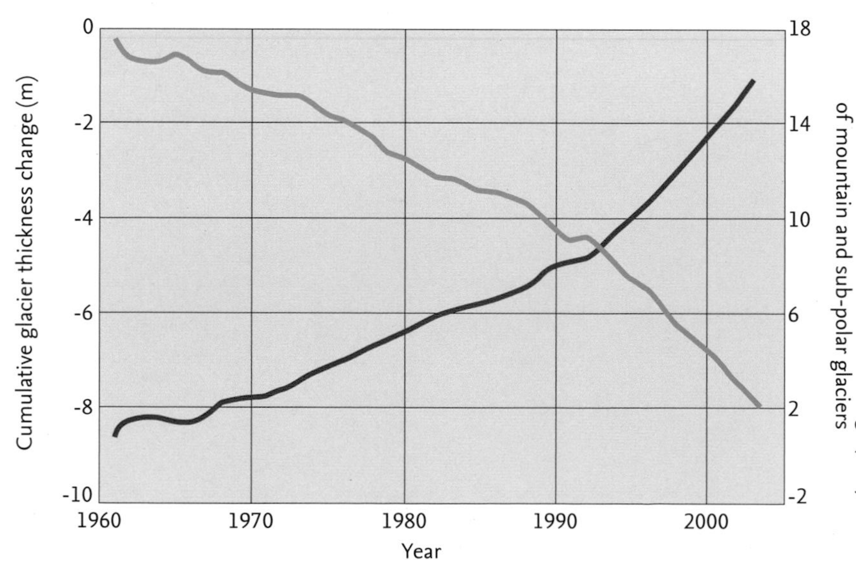

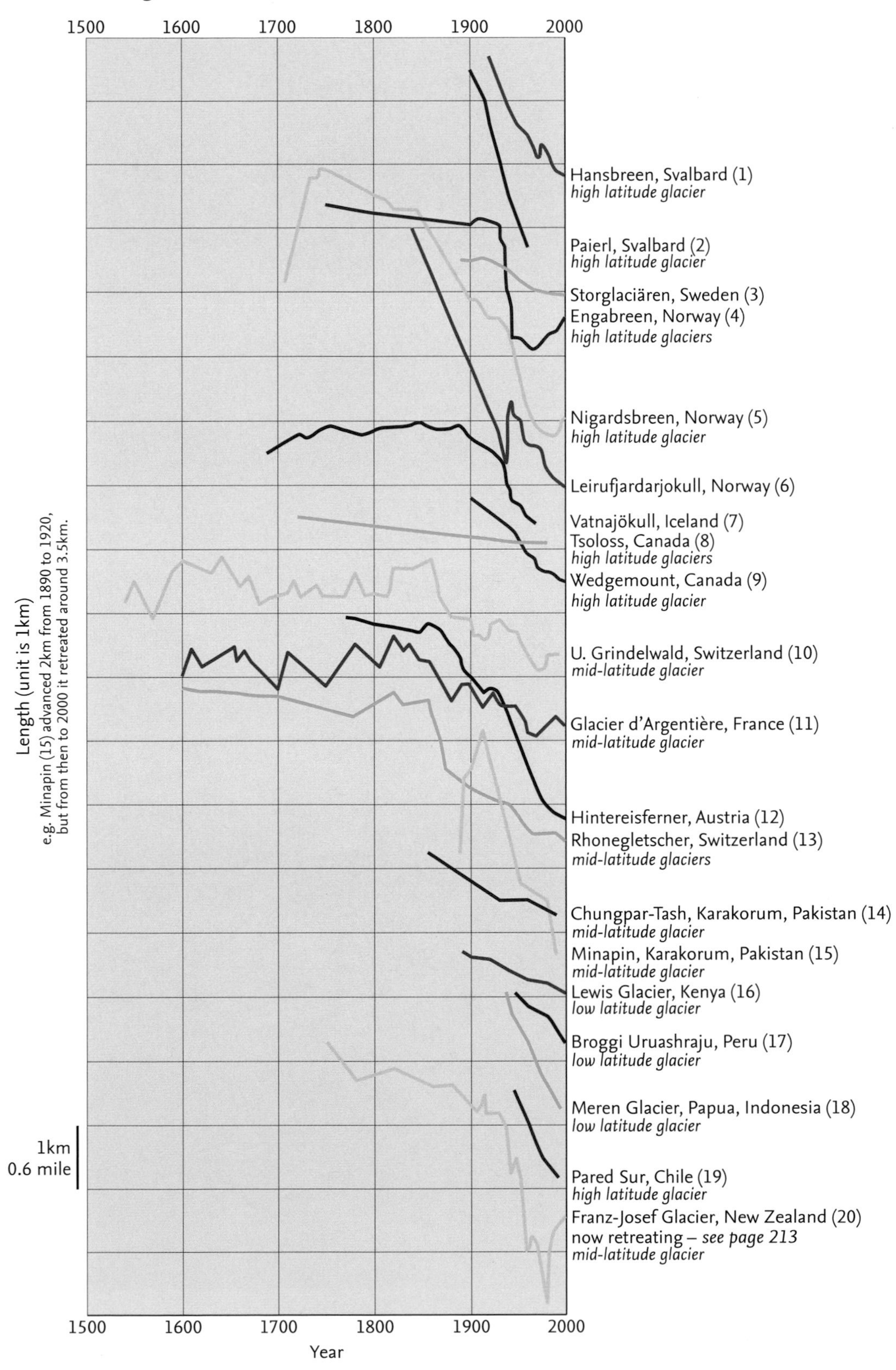

Monitoring of mass balance, the difference between accumulation and melting of a glacier, gives a measure of long-term patterns of advance and retreat. According to the World Glacier Monitoring Service 92 out of 113 glaciers worldwide saw a negative 'mass balance' for the year 2009–10, making it the twentieth consecutive year of negative balances. Glaciers with sustained negative balances will continue to retreat, and eventually disappear. The Chacaltaya glacier in Bolivia, once the world's highest ski run, melted at a faster rate than anticipated and finally disappeared in 2009.

A representative selection of twenty glacier length measurements from different parts of the world shows that glaciers are shrinking in countries as far apart as Norway and New Zealand. Retreat is widespread but the rates of retreat are most noticeable in mid-to-low latitudes, in ranges such as the Rocky Mountains and the Alps, as well as on isolated tropical summits. Over 110 glaciers have disappeared from Montana's Glacier National Park in 150 years, with scientists predicting a complete loss of glaciers from the park by 2030 if current trends continue. In the Alps nearly half the glaciers have disappeared since record keeping began, and scientists predict that the last of Africa's glaciers could be gone within twenty years.

RHONE GLACIER, SWITZERLAND 1900…

Glaciers are extremely sensitive to changes in climate, and most Alpine glaciers have retreated significantly during the last century, including the Rhone Glacier which is located in the Swiss Alps. The glacier has long been a popular tourist attraction and during the first half of the nineteenth century it used to terminate at the village of Gletsch at the bottom of a valley. The glacier is easily accessible due to its proximity to the Furka Pass road, and this has contributed to it being one of Switzerland's most studied glaciers with accurate observations running back to the 1870s.

...AND RETREATING 2008

Over the last 130 years the glacier has retreated by almost 1300 m (4265 feet), at an average rate of 8.5 m (28 feet) per year; its thickness has also reduced by 43 m (141 feet). The 2008 photograph of the glacier was taken from the same position as the tinted postcard image of it from 1900.

The glacier is no longer visible as it has retreated completely from the valley to a plateau which is not visible from this location. The 10 km-long glacier is predicted to continue its dramatic retreat during the twenty-first century and by 2100 there might not be anything left of it.

BLACKFOOT AND JACKSON GLACIERS, MONTANA, USA 1914 AND 2009

The Blackfoot Glacier in Glacier National Park was originally measured in 1850 at an area of 7.59 sq km (2.93 sq miles). With its reatreat it has separated into two parts as can be seen on the lower image, the Blackfoot Glacier on the left and and the Jackson Glacier on the right. Together they now cover 2.76 sq km (1.07 sq miles). These glaciers are part of the ongoing United States Geological Survey's Glacier Monitoring Research programme, which is researching changes to glaciers in and around Glacier National Park.

GRINNELL GLACIER, MONTANA, USA 1938 AND 2009

The Grinnell Glacier is one of the most photographed glaciers in the Glacier National Park. It is named after the early explorer George Grinnell. These photographs show that while the Grinnell Glacier has retreated significantly from the foreground over the intervening seventy years, the small Gem Glacier above it on a shelf in the mountain has hardly changed. But it has been predicted that all the glaciers in the park will have melted by 2030.

BRIKSDALSBREEN GLACIER RETREAT, NORWAY 2004 AND 2009

Briksdalsbreen is a northern arm of the Jostedalsbreen Glacier and it terminates in a small lake. It is a glacier which is affected by temperature and precipitation. It advanced in the early twentieth century then retreated mid-century exposing the lake. It was unusual in the 1990s as it advanced when other European glaciers were retreating. Since 2000 it has been retreating again and the terminus is now on land. It is thought this is its smallest extent since the 1200s.

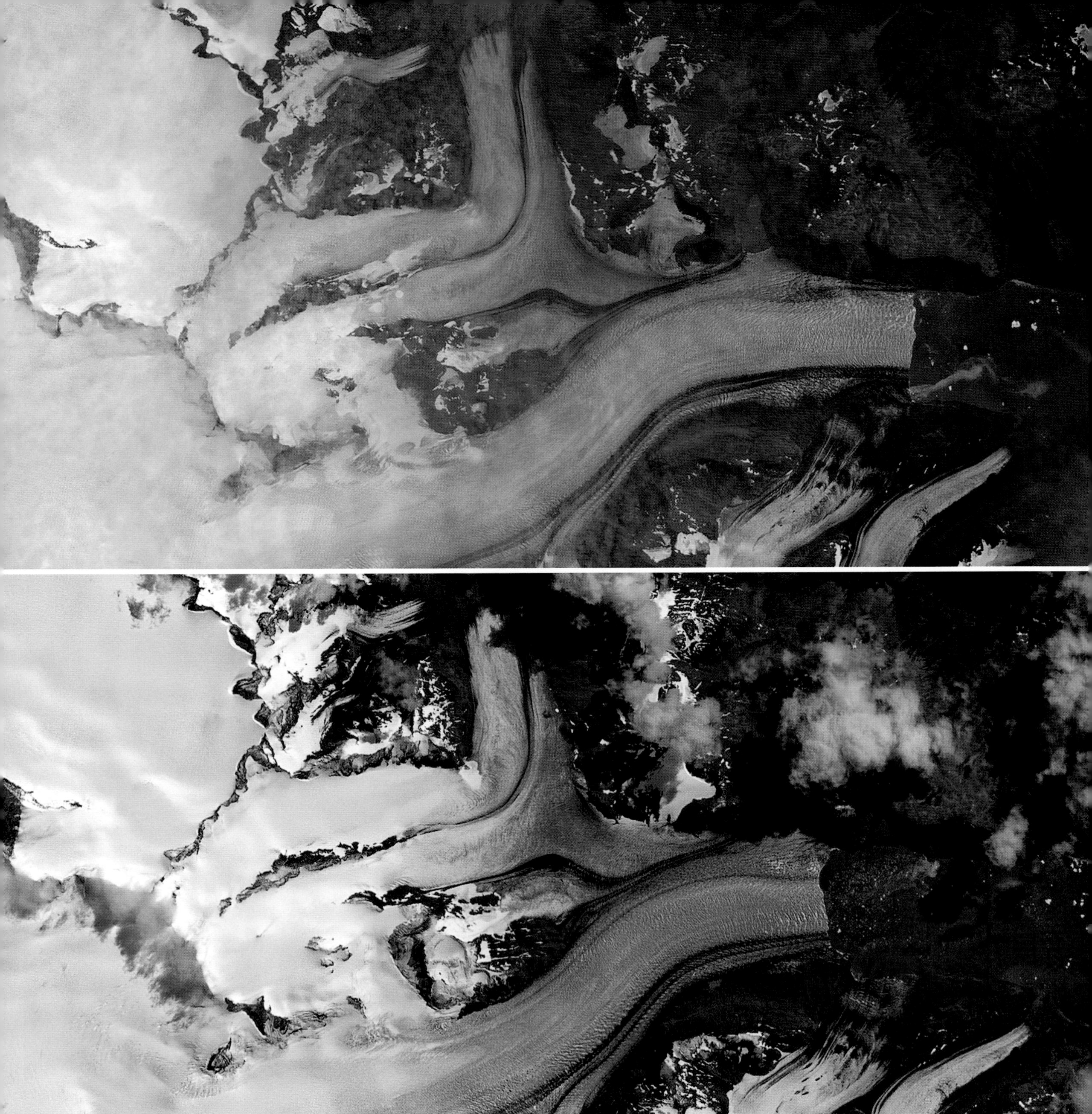

NEUMAYER GLACIER, SOUTH GEORGIA ISLAND 11 JANUARY 2005 AND 4 JANUARY 2009

Neumayer Glacier is located on the eastern coast of the island of South Georgia. Up until the 1970s, the tidewater glacier was retreating at a very slow rate of just tens of metres annually; however, between 1970 and 2002 the rate increased and the glacier retreated by 2 km (1.2 miles). Natural-colour satellite images captured the glacier's even more rapid retreat between 2005 and 2009, when it retreated approximately 1 km (0.6 miles). It is thought that a rise in sea temperature is having a negative effect on the glacier by causing its rate of flow to speed up and increasing its calving rate.

GANGOTRI GLACIER, INDIA 2001

This false-colour satellite image shows the Gangotri Glacier in northern India. At its current length of 30.2 km (18.7 miles) it is one of the longest in the Himalaya. The glacier is the source of the Bhagirathi river, an important tributary of the Ganges river. It is also a place of traditional Hindu pilgrimage. The Gangotri Glacier has been receding since 1780, and the retreat quickened after 1971. Over the last 25 years the glacier has retreated

more than 850 m (2789 feet), with a recession of 76 m (249 feet) from 1996 to 1999 alone. This is a concern as the glacial channel which feeds the river has changed course and the volume of water is shrinking rapidly, mainly due to reduced winter precipitation. Local deforestation around the glacier is also adding to the problem and it is feared that parts of the glacier may disintegrate.

FRANZ JOSEF GLACIER, NEW ZEALAND RETREATS, ADVANCES...

Franz Josef Glacier has shown a well-documented retreat over a number of years. The previously gradual retreat of the glacier was significantly accelerated by a number of warm winters and in this series of images it can be seen vanishing off into the distance in the middle of a period of rapid retreat which lasted over forty years. In the 1980s the snow conditions in this area changed. This reversed the retreat and the glacier advanced again, as can be seen in the 2001 image.

...THEN RETREATS AGAIN 18 MAY 2011

From 2010 the glacier started to retreat again. This very rapid retreat, which can be seen in this 2011 image, is also happening to other glaciers in the area. The Franz Josef Glacier has shown cycles of advance and retreat depending on the difference between the volume of meltwater at the foot of the glacier and volume of snowfall feeding the young granular snow or 'névé'.

RISING SEA LEVEL – average height of the level of the surface of the world's oceans and seas, likely to rise as a result of global warming

Camogli, Italy

RISING SEA LEVELS

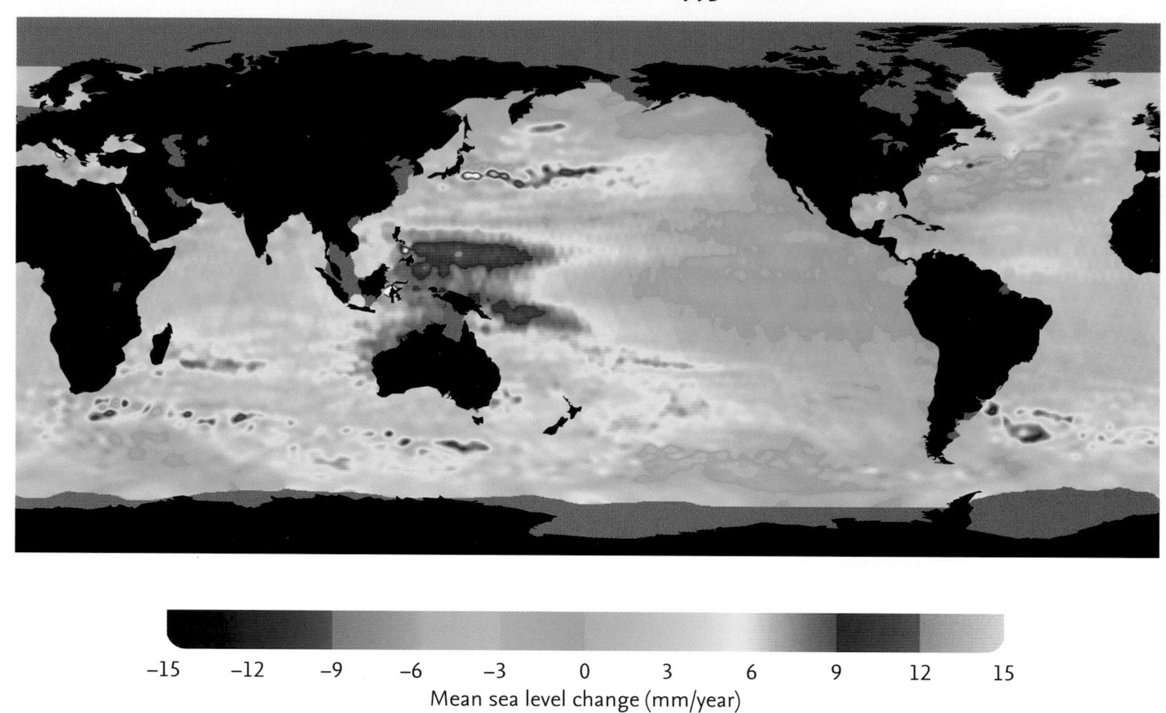

Most of the extra heat generated by the accumulation of greenhouse gases in the atmosphere goes to warm the oceans. This makes sea water expand in volume, raising sea levels around the globe. Most of the sea level rise which has so far been observed is due to this thermal expansion, but a significant contribution also now comes from shrinking glaciers and polar ice caps, which release fresh water as they melt.

Sea levels rose by between 12 and 22 cm (4.7 and 8.7 inches) during the twentieth century, and as global warming intensifies over the rest of the twenty-first century a continued rise is predicted. What is more difficult to predict is the exact magnitude of the sea level changes that lie ahead. The shrinking ice sheets of Greenland and Antarctica are undoubtedly a major contributory factor in the rise, but the complexity of glacier dynamics and the intricacies of modeling methods create wide-ranging predictions of change. The 2007 Intergovernmental Panel on Climate Change (IPCC) projection of a global sea level rise of between 18 and 59 cm (7 to 23 inches by 2090–99 has been challenged as too conservative by many, yet there is also research indicating that melt in Greenland is slower than previously thought and that sea level rises are unlikely to be as dramatic as worst-case scenarios have predicted. What is agreed is that ice melt is speeding up, and that with no sign of it slowing down there is undeniably a threat to low-lying coasts around the world, with even small rises likely to have a major impact in some places.

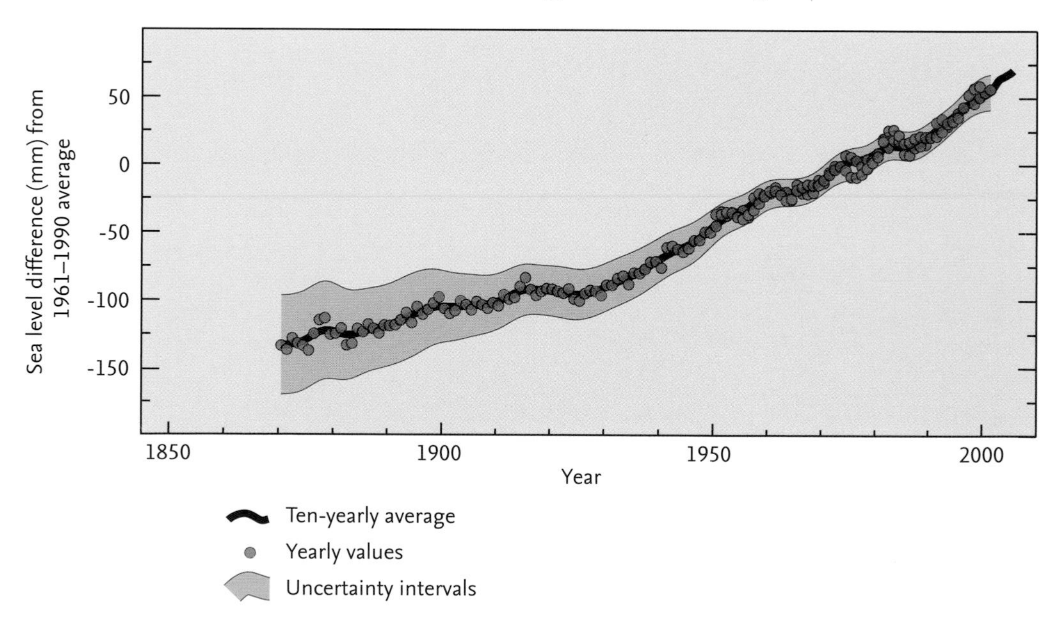

Changing sea levels are important because one-fifth of humanity lives within 30 km (18.6 miles) of the ocean. Major cities at risk from rising waters include Shanghai, New York, Mumbai and Tōkyō. It is predicted that many millions more people are likely to be flooded every year due to sea-level rise by the 2080s, with densely populated delta regions of Asia being the most vulnerable. In Bangladesh a one-metre rise in sea levels would flood 17 per cent of the country. Meanwhile, low-lying atolls such as the Pacific island nations of Tuvalu and Kiribati are also particularly vulnerable. The early twenty-first century heralded increasing fear that such low-lying nations would completely disappear beneath the sea in just a few decades. More recent study has show than many low-lying Pacific islands are growing due to coral debris and sediment build-up, rather than sinking due to sea-level rise, and the threat, although still existent and of great concern to inhabitants, is probably longer-term than first supposed.

Lowest Pacific islands

	Maximum height above sea level	Land area sq km	Land area sq miles	Population
Kingman Reef	1 m (3 ft)	1	0.4	0
Palmyra Atoll	2 m (7 ft)	12	5	0
Ashmore and Cartier Islands	3 m (10 ft)	5	2	0
Howland Island	3 m (10 ft)	2	1	0
Johnston Atoll	5 m (16 ft)	3	1	0
Tokelau	5 m (16 ft)	10	4	1 000
Tuvalu	5 m (16 ft)	25	10	10 000
Coral Sea Islands Territory	6 m (20 ft)	22	8	0
Wake Island	6 m (20 ft)	7	3	0
Jarvis Island	7 m (23 ft)	5	2	0

Threat of rising sea level

Major cities
Coastal areas at greatest risk
Islands and archipelagos
Areas of low-lying islands

VENICE, ITALY

The city of Venice covers 117 islands in the saltwater Venetian Lagoon. The buildings are constructed on closely spaced wooden piles with the foundations of the brick or stone buildings on top. Today the city is one of the top tourist destinations in the world. Among its main sights is St Mark's Basilica, one of the best known examples of Byzantine architecture, built in 832 and now the cathedral church of the Roman Catholic Archdiocese of Venice.

HIGH WATER IN VENICE, ITALY

In winter Venice experiences a few exceptional tides, the 'acqua alta'. This is due to a combination of factors such as astronomical tide, strong south winds, some subsidence of the city over time and the Adriatic 'seiche' wave movement.

Lower-lying parts of the city such as St Mark's Square are most often affected, but only for a few hours until the tide ebbs. Visitors make use of the 'passarele' walkway to visit the basilica when this happens.

VENETIAN LAGOON, ITALY 9 DECEMBER 2001

At 550 sq km (212 sq miles), the Venetian Lagoon is Italy's largest wetland. The coastal lagoon is comprised of hundreds of islands, and vast areas of mudflats, tidal shallows and salt marshes. It is connected to the Adriatic Sea by three inlets, and the tidal influence ensures that there are large fluctuations in water levels within the lagoon. On occasion the built-up areas around the lagoon suffer severe floods. To minimize the impact, flood-defence schemes have been put in place, including reinforcement of sea walls and jetties, and beach and sand dune reconstruction.

FLOOD DEFENCES IN VENICE, ITALY 22 JUNE 2008 AND 2009

Sea level rise is a major threat to the low-lying city of Venice. The city is also experiencing an increase in the frequency of its seasonal high-tide flooding events (acqua alta). The MOSE Project, scheduled to be completed in 2014, is a system of seventy-eight mobile flood barriers situated across the three inlets to the Venetian Lagoon. Whenever the Adriatic Sea rises 1.1 m (3.6 feet) above average, the barriers will be raised to prevent sea water from surging into the lagoon and flooding the city. The second image shows the construction of the forty-one flood barriers at the Lido inlet.

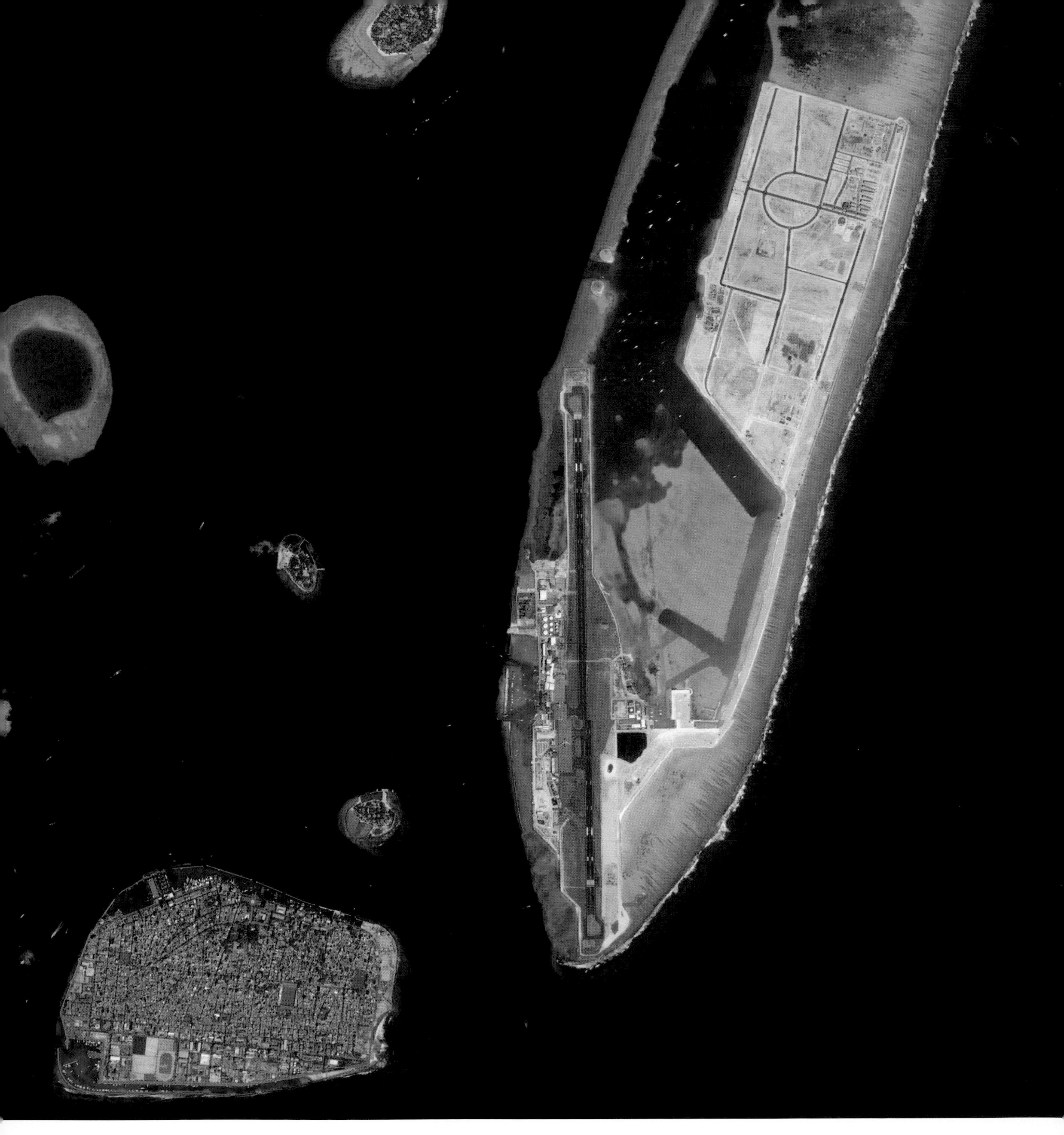

THE MALDIVES, INDIAN OCEAN...

In the Maldives, in the Indian Ocean, storms do not raise the water level by more than around 30 cm (11.8 inches). Accordingly, development has been concentrated on land only 40 cm (15.7 inches) or more above sea level. This means that they are extremely vulnerable to sea level rise. Most of the population lives within 2 m (6.5 feet) of sea level and almost all within 4 m (13.1 feet).

...LOW-LYING AND VULNERABLE TO SEA LEVEL RISE

Male, the capital of the Maldives, is approximately 2 m (6.5 feet) above the sea, but it has reclaimed areas of land which is lower. After storms in 1987 and 1988 flooded the reclaimed areas, a series of breakwaters on the outer coast were built to protect the town from damaging storm waves, but they will not prevent flooding from a sustained rise in sea level.

THE CLIMATE CHALLENGE

Mark Maslin

The World is warming up and the big thaw has already started. For the twentieth century we have clear evidence for a 0.75°C (1.35°F) rise in global temperatures and a 22 cm (8.7 inches) rise in sea level. We have seen significant shifts in the pattern of the seasons, changing weather patterns, and significant retreat of Arctic sea ice and nearly all continental glaciers. According to the US National Aeronautics and Space Administration (NASA), the UK Met Office and the Japanese Meteorological Agency, the last decade has been the warmest on record. The Intergovernmental Panel on Climate Change (IPCC) predicts that by 2100 global temperatures could rise by between 1.8°C and 4.0°C (3.2°F and 7.2°F), depending on the levels of greenhouse gas emissions over the next ninety years. Sea level could rise by between 30 cm and 80 cm (11.8 and 31.5 inches), even more if the melting of ice in Greenland and Antarctica accelerates. Weather patterns will become less predictable and the occurrence of extreme climate events, such as storms, floods, heat waves and droughts, will increase.

The temperature of the Earth is determined by the balance between energy from the Sun and its loss back into space. Most of the incoming solar short-wave radiation (ultraviolet and visible 'light') passes through the atmosphere without interference. About one-third of the solar energy is reflected straight back into space, while the remaining energy is absorbed by both the land and ocean, and is released as long-wave infrared radiation or 'heat'. Greenhouse gases such as water vapour, carbon dioxide, methane, and nitrous oxide can absorb some of this long-wave radiation, thus warming the atmosphere. We need this 'greenhouse effect' because without it, the Earth would be at least 35°C colder, making the average temperature in the tropics about -5°C. Since the industrial revolution we have been burning fossil fuels (oil, coal, natural gas), releasing carbon back into the atmosphere, increasing the effect – we have already increased the amount of carbon dioxide in the atmosphere by over 40 per cent. In effect every time we switch on a light or start our car we are releasing ancient stored solar energy back into the climate system and warming the Earth. In one century we have caused changes on a scale that would have taken thousands of years through purely natural processes.

The impacts of global warming will increase significantly as the temperature of the planet rises. The frequency and severity of floods, droughts, heat waves and storms will increase and coastal cities and towns will be especially vulnerable as sea-level rise will affect the magnitude of floods and storm surges. The increase of extreme climate events coupled with reduced water-security and food-security will have a severe effect on the health and wellbeing of billions of people.

So how far can we let this big thaw continue until we take serious action to slow it down? Politicians and scientists seem to agree that a limit of a rise of 2°C above pre-industrial average temperatures is needed. Below this threshold it seems that there were both winners and losers due to regional climate change, but above this figure everyone seems to lose. However, with the failure to produce a new climate treaty it now seems likely that temperature rise will exceed this threshold. This is despite the fact that the Stern Review in 2007 suggested that dealing with climate change would only cost us 1–2 per cent of world GDP every year, while the impacts could cost between 5 and 20 per cent.

Climate change is the major challenge for our global society. Despite all the benefits that fossil fuels have brought us we now need to replace them with sustainable carbon-free renewable or alternative energy sources. This is essential, as the power needs of humanity will continue to expand due to the rapid development of China, India, and other countries. Sophisticated political solutions are required at all levels; ranging from a binding international agreement, to regional, to national, and to local policies. For example the European Emissions Trading Scheme is driving down EU emissions, while the UN Clean Development Mechanism tries to promote low carbon investment in developing countries. At national level, the UK and Mexico are leading the world with binding climate change laws. There is also a need for massive investment in sustainable energy generation and low carbon technology, to provide the means of reducing world carbon emissions. The scientific evidence for climate change is unequivocal and it is due to us. But cutting global greenhouse emissions need not be such a daunting task and many of the solutions are win-win, with increased health benefits and energy security.

Dramatic evidence of glacier retreat is shown in this photograph of the Athabasca Glacier, Canada – North America's most visited glacier. It has lost half its volume and has receded by nearly a mile in the last 125 years.

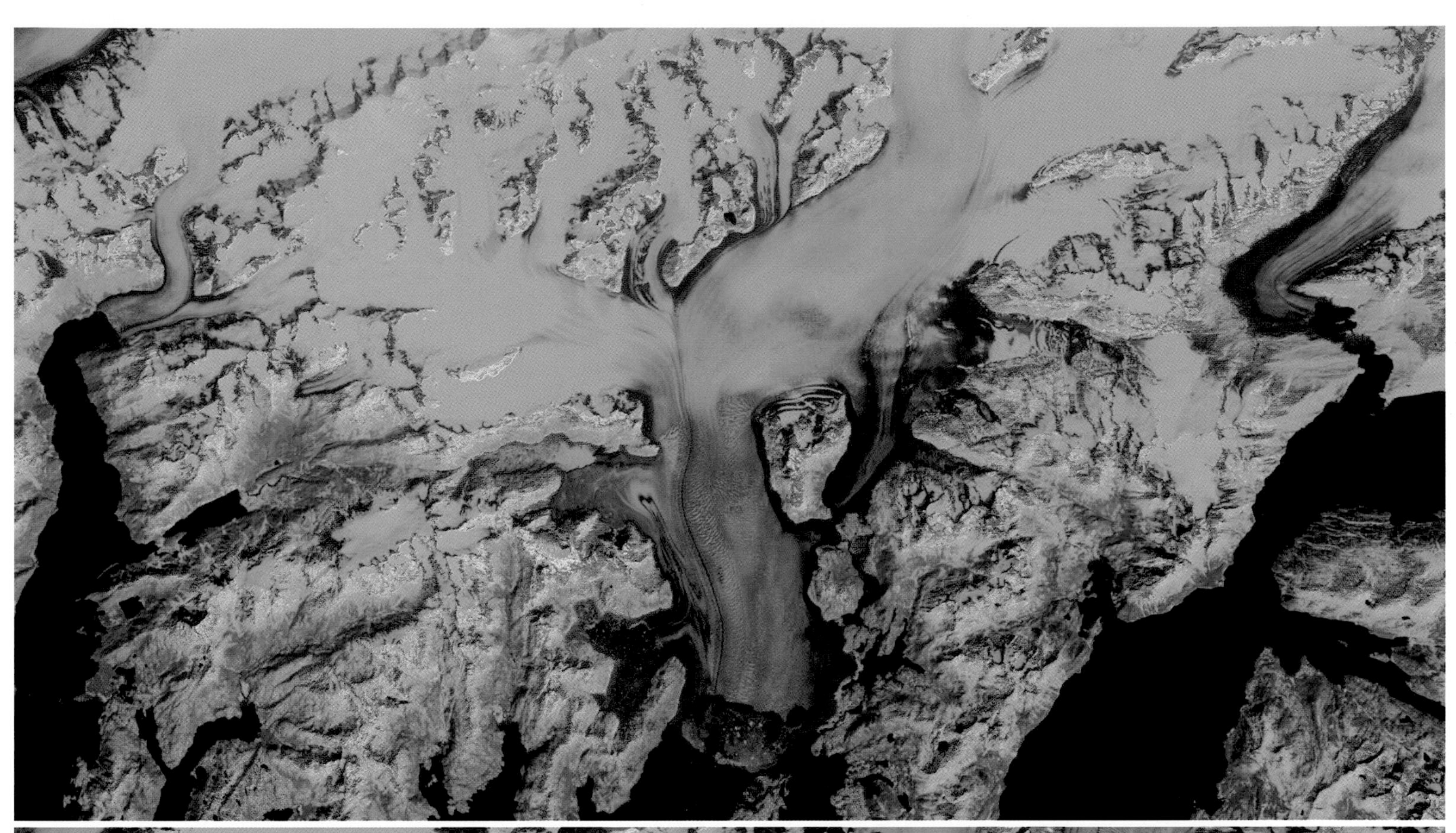

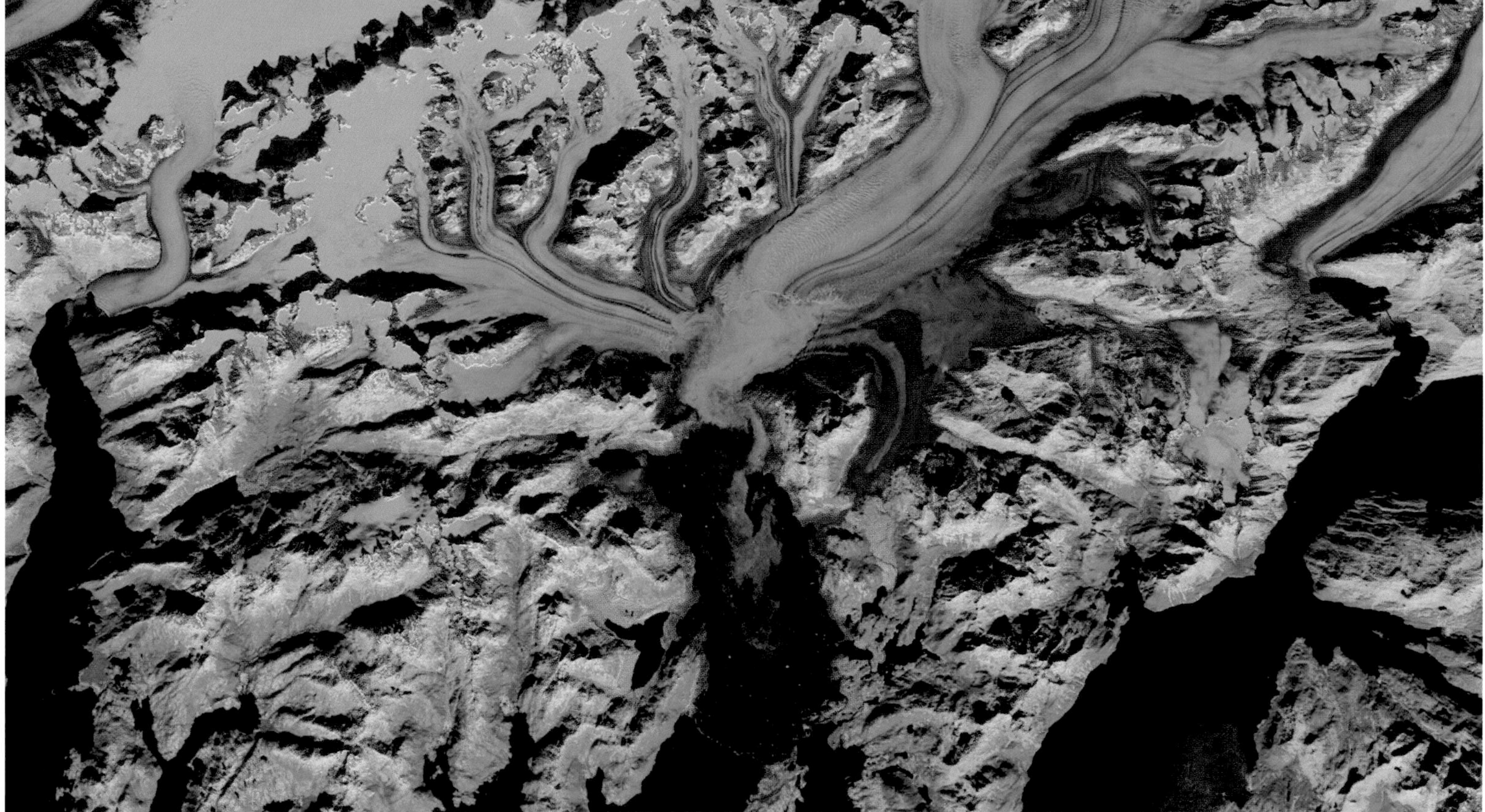

False colour satellite images showing the retreat of the Columbia Glacier, Alaska, USA between 1986 and 2010. The glacier has retreated by more than 19 km (12 miles) and its thickness has reduced significantly, as evidenced by the growing areas of bedrock (red-brown).

PARCHED

EARTH

DESERTS AND DROUGHT

Advancing deserts – the encroachment of desert conditions into settlements or agricultural areas as a result of climate change or bad farming practices

Drought and fire – prolonged period of dry weather leading to water shortages, loss of crops and an increased risk of fire

Shrinking lakes, drying rivers – reduction in the size of lakes and river flow rates due to climate change or the extraction of water for agriculture or industry

ADVANCING DESERTS – the encroachment of desert conditions into settlements or agricultural areas as a result of climate change or bad farming practices

Namib Desert, Namibia

ADVANCING DESERTS

Each spring, billowing clouds of dust sweep down from the drylands in the north of China and envelop Beijing. Homes, streets and cars are left coated with brown dirt, and the skies turn an opaque orange colour. Sand dunes are now a mere 70 km (43.5 miles) from Beijing itself, and huge areas of Inner Mongolia, Gansu and Xinjiang provinces are being engulfed by expanding desert.

But the Chinese are fighting back. Efforts have been made to stop the overgrazing by sheep and goats which triggers desertification by removing grass and bushes, which lets strong winds erode the sandy soil. Ploughing has also been banned in areas where it allows topsoil to blow away. The Chinese government is also planting a 'green wall' of trees – almost as long as the famed Great Wall of China – to try to stop the advancing sands. It claims that anti-desertification measures have slowed the rate of desertification from 3400 sq km (1313 sq miles) to 1200 sq km (463 sq miles) per year.

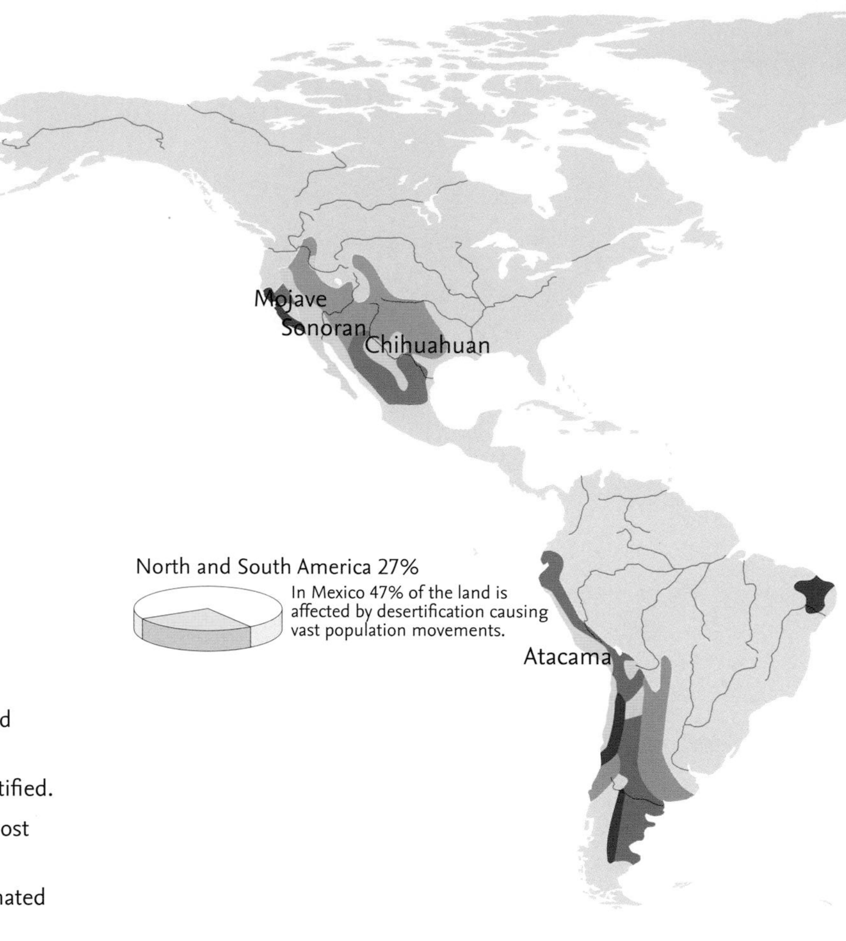

- Land degradation affects more than 1 billion people throughout the world and 40 per cent of the earth's surface.
- About 36 million sq km (13.9 million sq miles) of the earth's land is desertified.
- Every year about 120 000 sq km (46 300 sq miles) of land worldwide are lost to degradation and the rate is increasing.
- Desertification threatens people's livelihoods and contributes to an estimated US$42 billion in lost incomes.
- Desertification is a contributory factor in the internal displacement and international migration of people.
- It is predicted that by the 2050s, 50 per cent of agricultural land in Latin America will be subject to desertification.

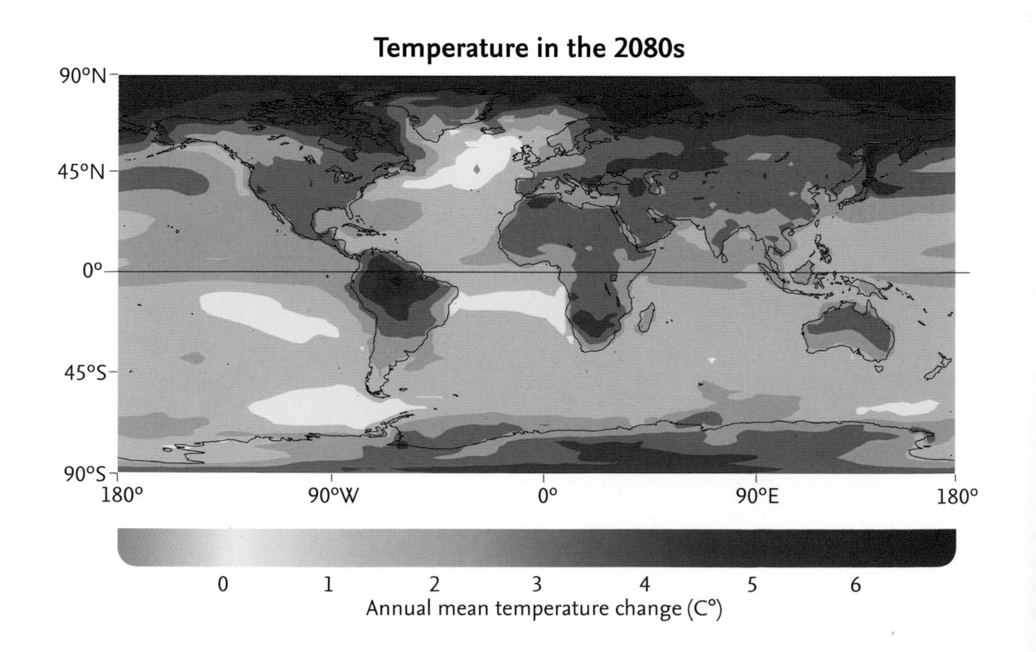

Land at risk of desertification

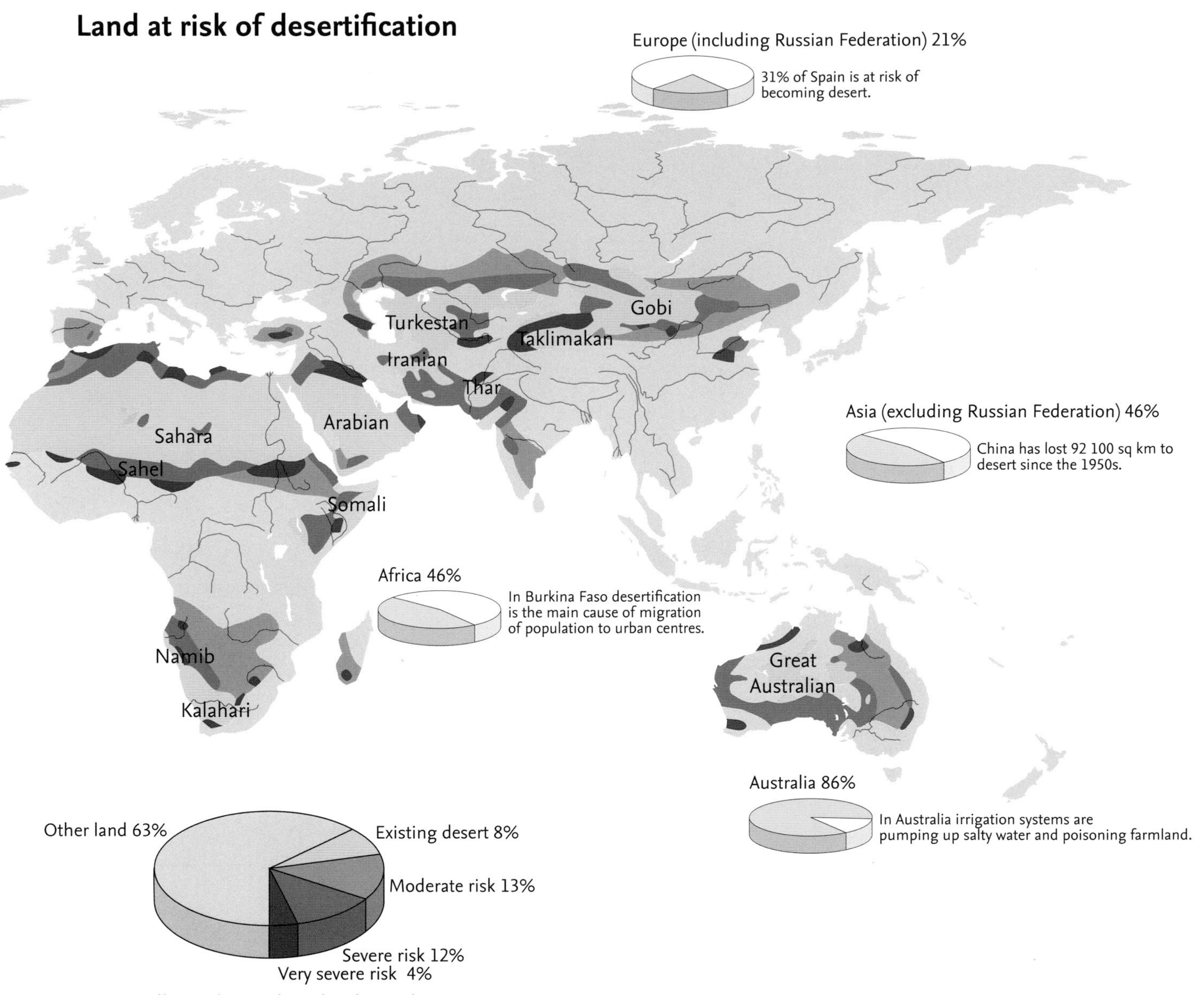

Small pie charts show land at risk of desertification by continent

Precipitation in the 2080s

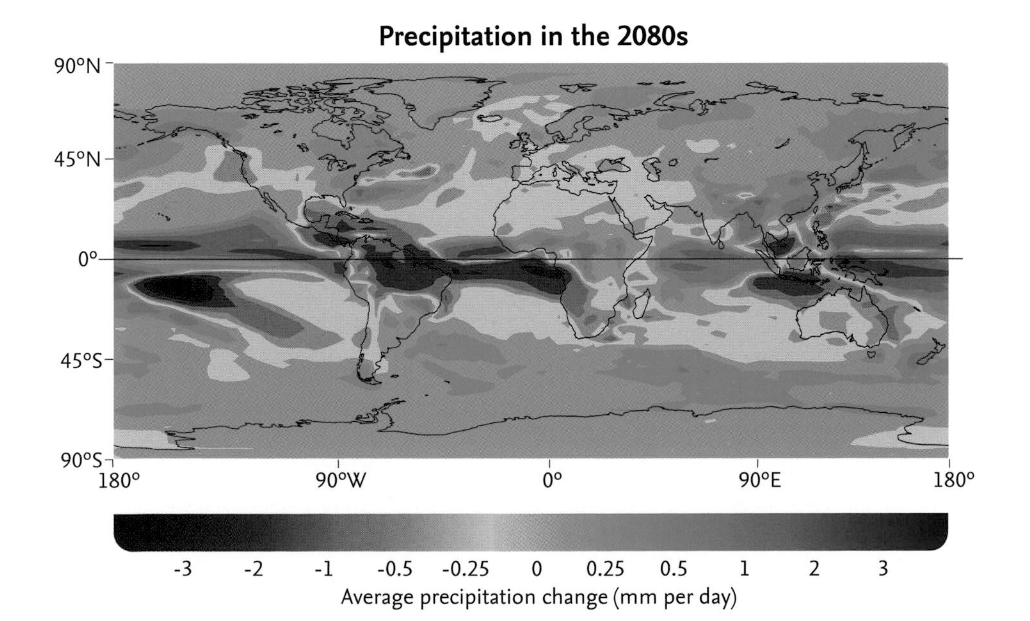

China is not the only area affected. Desertification is a major issue for drylands and desert areas across the world, particularly the Sahel in Africa (on the southern fringe of the Sahara), arid areas of central Asia and the southern Kalahari fringe of Africa. Even parts of Europe – particularly central Spain – are now affected by desertification, with the situation aggravated by drought.

The incidence of drought seems to be increasing across the world, probably because of global warming. Scientific studies have linked droughts in Africa and the United States to rising ocean temperatures which change weather patterns in complex ways. Computer model projections for climate change over the next century suggest that this problem will get dramatically worse – as temperatures rise further, new deserts will spread across southern Africa, central North America and the Mediterranean fringe of Europe.

THE EDGE OF THE DESERT, SAHEL REGION, AFRICA

Many parts of the world, on the delicate fringes between fertile and arid regions, are at risk from advancing deserts. This is shown graphically in this image of the Sahel region of Africa. The Sahel is a buffer zone between the Sahara desert to the north and the savannah grasslands to the south. Sand dunes are also threatening oases in the desert and various measures are being taken to try to restrict the advance of sand onto valuable arable

and cattle-rearing land. Climate change and man's destruction of natural vegetation are exacerbating the problem and threatening the traditional way of life in many areas.

DESERT GARDEN, ALGERIA 2009

One of the countries most affected by desertification is Algeria in northern Africa. The Sahara desert covers much of central and southern Algeria, and the arid climate is a challenge to arable farming. Scarce and infrequent water supplies coupled with soils that tend to erode easily, ensure that only 3 per cent of the country's land is cultivated, most of which is located along the fertile northern coastal areas. Algeria is experiencing rapid population growth, which is putting increased pressure on the Sahara's already limited water supplies.

The aerial photograph is of a garden in part of a village located near a naturally occurring spring, Aïn Hammou, in the Algerian Sahara. The garden is nestled in a hollow between the sand dunes; the outer ridge is topped with rows of palm trees to stabilize the dunes and prevent the sand from burying the plants within the garden. Its existence is due to the supply of groundwater in the area; traditionally water would be hauled up from hand-dug wells using pulleys and ropes, however, electric pumps now make the task easier.

EGYPT AND THE NILE 19 JULY 2004

The Nile Valley in Egypt illustrates the fine line between desert and fertile land. Completely surrounded by desert, the Nile is a lifeline for the Egyptian people. Until the Aswan Dam was built the Egyptians relied on the annual flooding of the Nile to irrigate their land but this flooding was not guaranteed and drought and famine were commonplace.

SAHEL, NEAR AKABLI, ALGERIA 2009

An efficient and sustainable desert irrigation system exists in the village of Sahel, near the town of Akabli, deep in the Algerian Sahara. Each farmer is allowed a certain quota of water depending on his inheritance rights. The village is located in an oasis that receives water from a foggara (man-made underground aqueduct). In order for a foggara system to work, the oasis must be located downhill of a groundwater source so that the force of gravity pushes the water downwards through underground pipes. This ancient system provides villages with fresh, reliable supplies of water.

FIGHTING DESERTIFICATION IN MAURITANIA, AFRICA JANUARY 2010

There are many ways of controlling advancing deserts, such as planting trees and nitrogen-fixing plants, or spraying petroleum over seeded land to prevent moisture loss. More simply, as in the photo above, brushwood fences are constructed along sand dunes to stabilize them and stop them further encroaching on arable land or human settlements. The fences also act as windbreaks. Mauritania employs a combination of desertification control methods to ensure that the country's small amount of arable and pastoral land does not diminish any further.

PROTECTING THE TARIM DESERT HIGHWAY, CHINA 1 MARCH 2009

The Taklimakan Desert, one of the world's largest deserts, covers an area of 337 000 sq km (130 116 sq miles); 85 per cent of its surface is made up of huge shifting sand dunes, which are propelled by strong winds. The 552-km (343-mile) Tarim Desert Highway crosses the desert from north to south, and was built in the 1990s by the Chinese government to gain access to untapped oil and gas fields. Grass grids and millions of trees, fed by extensive irrigation networks, were planted by the roadside to prevent it from becoming covered by advancing sand dunes.

DROUGHT AND FIRE –

prolonged period of dry weather leading to water shortages, loss of crops and an increased risk of fire

Alvord Desert, Oregon, USA

DROUGHT AND FIRE

Major droughts 1964–2012

Year	Country	Deaths
2011-12	Somalia	na
2011-12	Djibouti	na
2011-12	Ethiopia	na
2011-12	Kenya	na
2006	China	134
2005-06	Burundi	120
2005-06	Kenya	27
2004	Kenya	80
2002	Malawi	500
2002	Uganda	79
2001	Angola	58
2001	Guatemala	41
2000-01	India	20
2000-01	Somalia	21
2000	Afghanistan	37
1999-2003	Pakistan	143
1999-2002	Kenya	85
1999-2001	Uganda	115
1997-98	Papua New Guinea	60
1997	Indonesia	672
1991	China	2 000
1989	Rwanda	237
1988-92	Madagascar	200
1988	China	1 400
1987	Somalia	600
1987	Ethiopia	367
1987	India	300
1987	Mozambique	50
1986	Indonesia	84
1984	Indonesia	230
1983-85	Sudan	150 000
1983-84	Ethiopia	300 000
1983	Swaziland	500
1983	Brazil	20
1982	Indonesia	280
1981-85	Mozambique	100 000
1978	Indonesia	63
1974-76	Somalia	19 000
1973-78	Ethiopia	100 000
1967-69	Australia	600
1966	Indonesia	8 000
1965-67	India	1 500 000
1965	Ethiopia	2 000
1964	Somalia	50

Fire is a natural part of the ecosystem in many parts of the world. In Australia, many species of eucalyptus trees require their seed pods to be scorched by fire before they will open and germinate. Australia's Aboriginal people used 'firestick farming' to clear grasslands and manage their hunting grounds.

Fire does not sit well with heavily-populated urban areas, however, where millions of dollars' worth of real estate – not to mention a family's entire possessions – can go up in flames once a wildfire gets out of control. The worst fires can also be deadly: in 1983 twelve firecrew members were killed near Melbourne in a feared 'crown fire' which jumped between tree tops. In 1994 Sydney was almost encircled by over 800 separate fires, which rained ash onto the central business district and shut out the sun as brown smoke drifted over vast areas.

Fire and drought are closely linked. The worst fires of the modern era struck the forests of Indonesia during the drought sparked by El Niño in 1998. The conflagrations not only blotted out entire countries with smoke and smog, causing respiratory diseases and grounding aircraft, but released millions of tonnes of carbon dioxide from burning underground peat and above-ground vegetation, worsening climate change.

Humans are directly involved in this process. The Indonesian fires were partly caused by farmers clearing land for palm oil plantations. The 2009 wildfires in Victoria, Australia occurred during a period of severe drought and a sweltering heatwave, but were sparked by various causes, such as fallen power lines, carelessness and arson. When the flames finally died down, 173 people were dead and 414 injured.

There is little doubt that global warming will make fires worse. The Mediterranean area can expect the burning season to lengthen by several weeks. Australia and California will also be severely affected, while areas in the far north such as Alaska – where forest fires are normally a rarity – will also see huge areas turned to ashes.

Major wildfires 1967–2011 and droughts 1964–2012

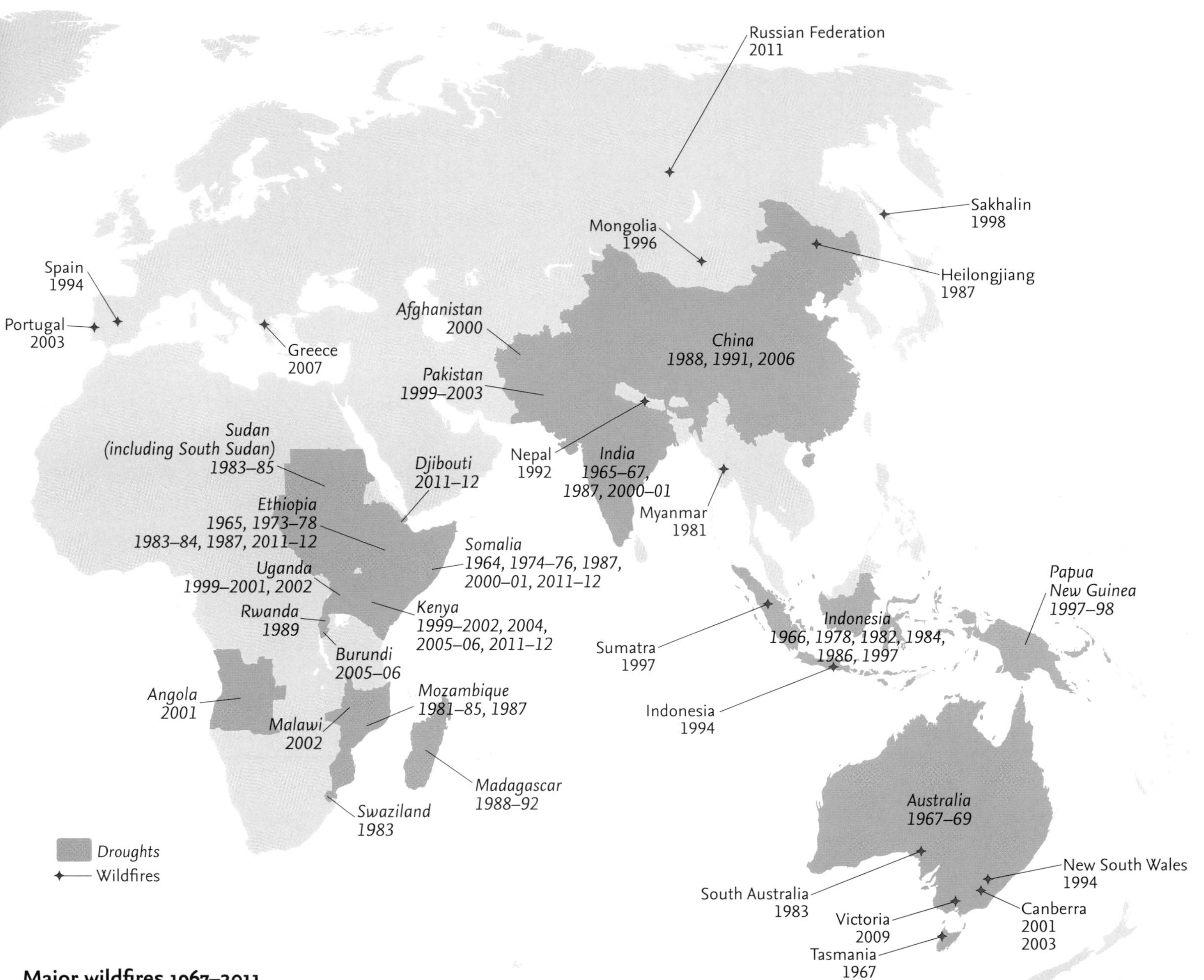

Major wildfires 1967–2011

Year	Location	People affected	Area affected sq km	sq miles
2011	Russian Federation	20 000	6 356	2 454
2009	Victoria, Australia	9 954	4 500	1737
2007	California, USA	640 064	2 000	772
2007	Greece	5 392	2 700	1 042
2003	Canberra, Australia	2 650	30 000	11 583
2003	Portugal	150 000	4 200	1 622
2003	California, USA	27 104	1 133	437
2001	Canberra, Australia	4 400	5 500	2 124
2000	Arizona/New Mexico, USA	26 400	680 194	262 625
1998	Roraima, Brazil	12 000	9 254	3 573
1998	Sakhalin, Russian Federation	100 683	5 000	1 931
1998	Florida, USA	40 124	1 052	406

Year	Location	People affected	Area affected sq km	sq miles
1997	Sumatra, Indonesia	32 000	800	309
1996	Mongolia	5 061	80 000	30 888
1994	California, USA	1 200	10 000	3 861
1994	New South Wales, Australia	26 020	8 000	3 089
1994	Spain	15 020	2 700	1 043
1994	Indonesia	3 000 000	1 360	525
1992	Nepal	50 000	unknown	
1987	Heilongjiang, China	56 313	25 000	9 653
1987	Argentina	152 752	unknown	
1983	South Australia, Australia	11 000	unknown	
1981	Myanmar	48 588	unknown	
1967	Tasmania, Australia	3 100	unknown	

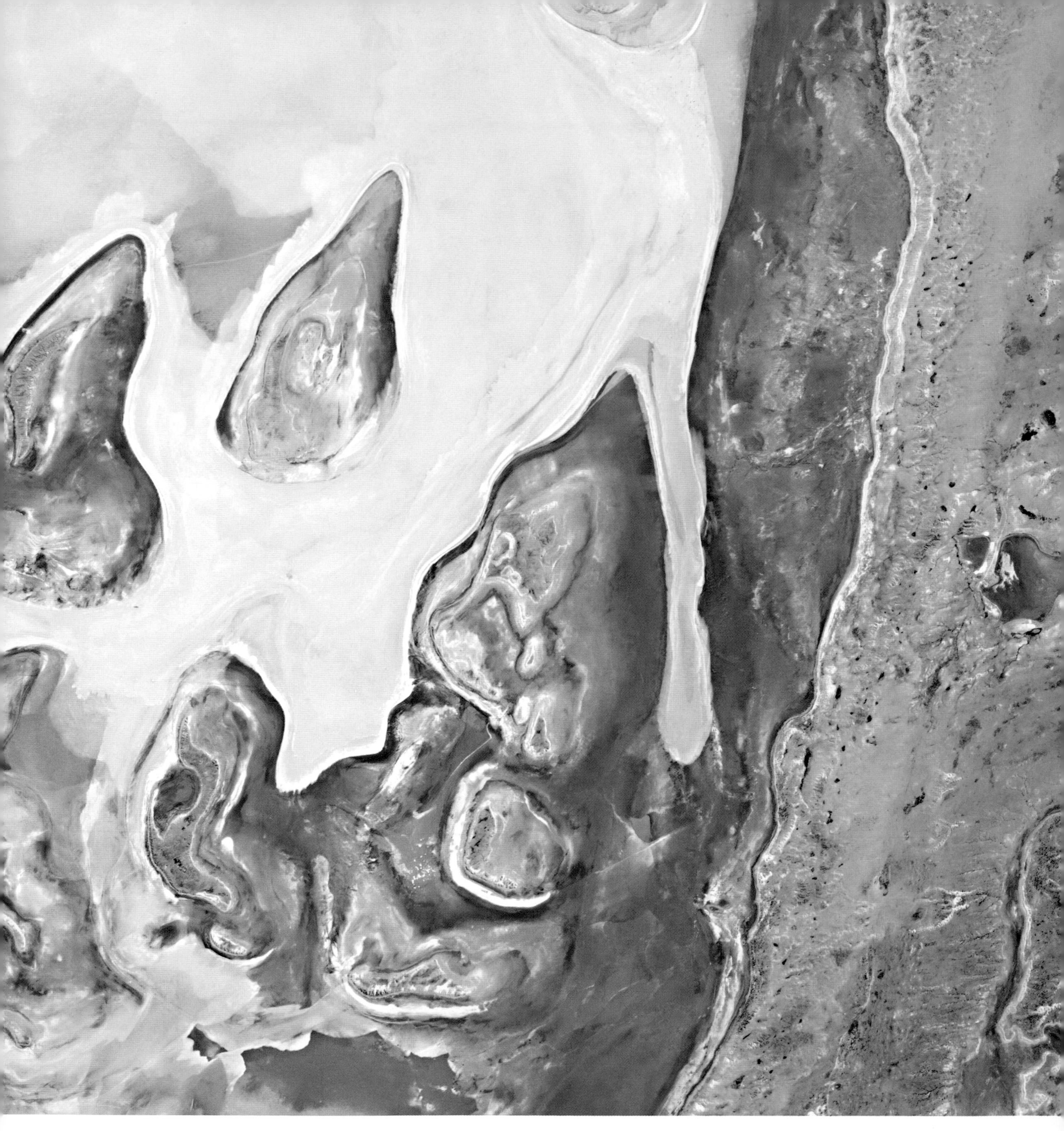

LAKE FROME, AUSTRALIA 2009

Lake Frome is the most southerly of the lakes which make up an endorheic, or closed, drainage basin in the area to the east of the Flinders Ranges. As this area receives very little rain in an average year these lakes are usually saltpans. This satellite image of the eastern edge of the lake shows the normal view of dry river channels, with the uneven surface of the saltpan filled with dry sediment.

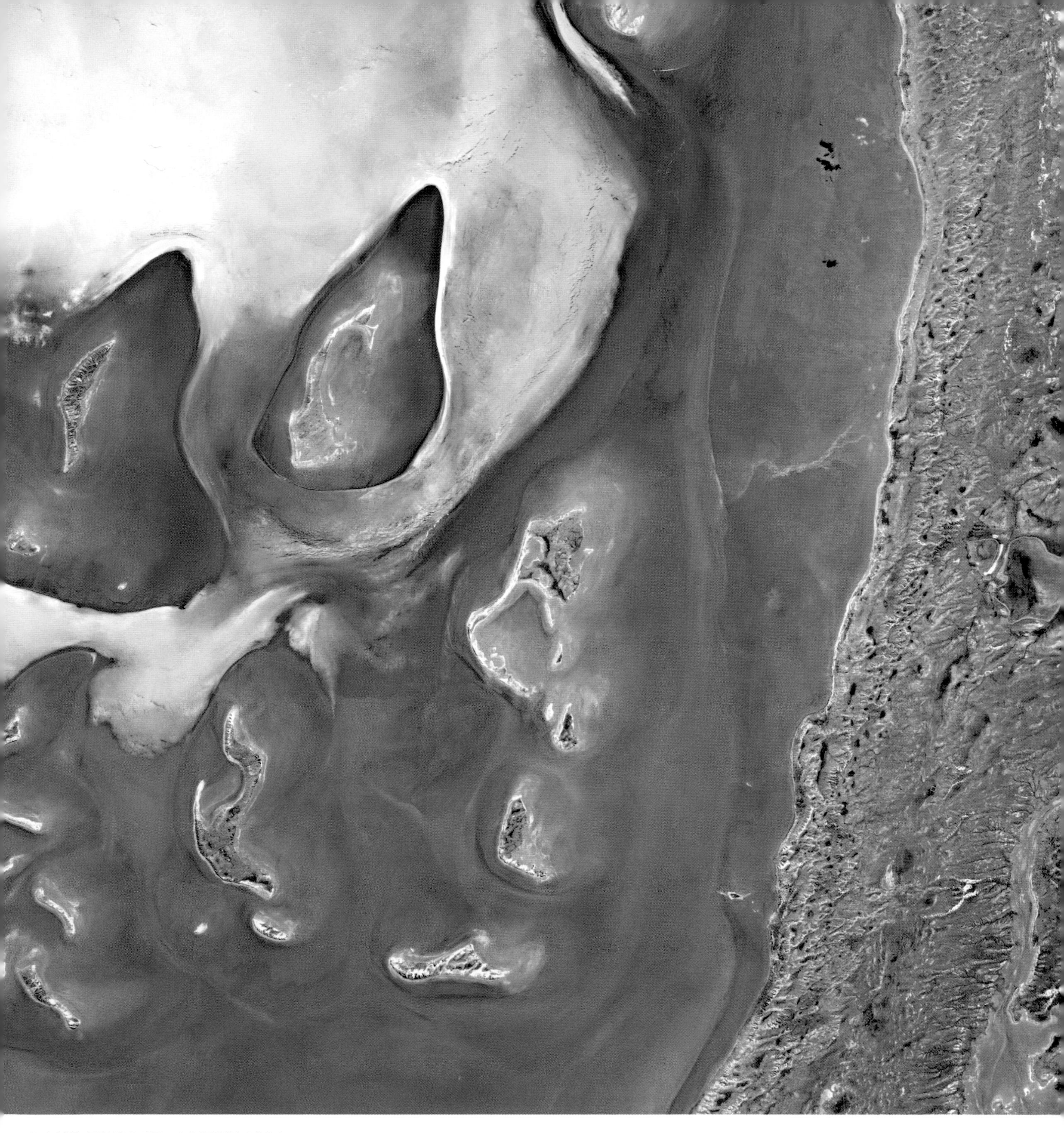

LAKE FROME, AUSTRALIA 2010

Some rains do make it into Lake Frome as overflow from Lake Callabonna to the north or from rain in the northern Flinders Ranges. In early 2010 rains in central Australia caused severe flooding with impermanent rivers being filled all the way from Queensland to South Australia. Gradually the water travelled southwards and eventually some made its way to Lake Frome, which was one of the last areas to fill with water, where it appears green as a muddy 'river'.

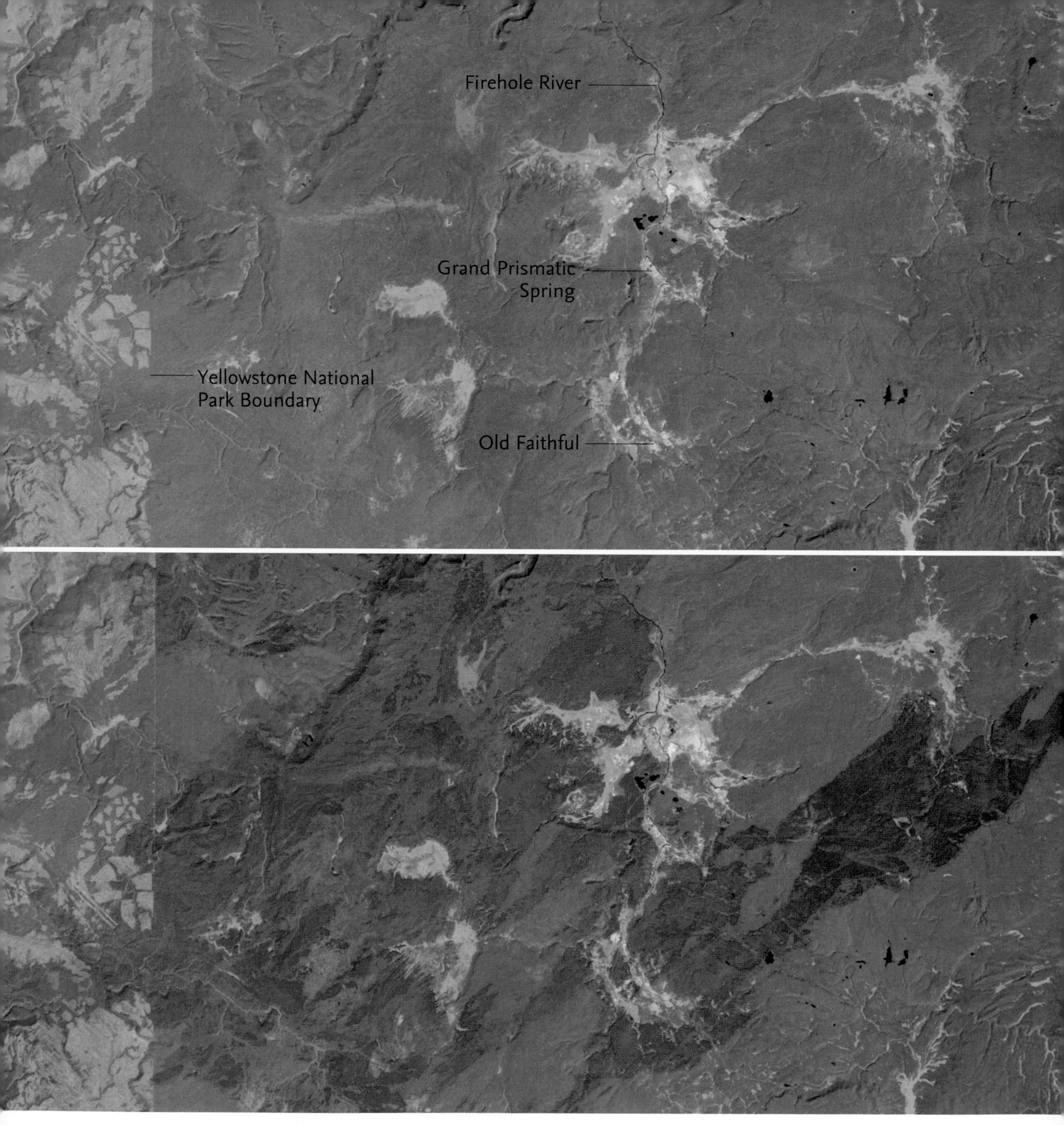

YELLOWSTONE NATIONAL PARK, USA 1987 AND 1989

During the hot, dry summer of 1988 thunderstorms crossed Yellowstone National Park and the associated lightning set fire to the dry vegetation. By the end of the summer there had been 50 wildfires and approximately 35 per cent of the park had been affected. In the 1987 image there is dense mature pine forest showing dark green, with light green meadow, light blue geyser fields and dark blue lakes. In the 1989 image, the burn scars are dark red and show a patchwork of fire damage and the full extent of the fire damage can still be seen. There are burn scars all around the area

YELLOWSTONE NATIONAL PARK, USA 1998 AND 2011

of Old Faithful in the bottom right, but not everything was damaged. Fire helps the lodgepole pine and grassy meadows of the park to regenerate, but the environment means that regeneration is slow and scarring is still clear in 1998, especially on the high plateau areas where the growing season is short. Twenty-three years on from the fire the 2011 image shows the forest is returning and the burn scar is fading as young trees grow.

HEAT WAVE AND WILDFIRES IN RUSSIA 5 AUGUST 2010

In July and August 2010, record-breaking summer temperatures led to the worst heat wave in modern Russian history, with more than 55 000 deaths due to heat stress and respiratory illness. The combination of high temperatures and associated drought, led to an outbreak of 850 wildfires across central and western Russia. Despite the best efforts of 180 000 firefighters and countless volunteers, forest and peat-bog fires destroyed hundreds of thousands of kilometres of countryside; 52 people were killed, and 3000 people were left homeless.

ST BASIL'S CATHEDRAL, MOSCOW, RUSSIA BEFORE THE WILDFIRE SMOG AND ON 7 AUGUST 2010

Moscow, Russia's capital city, was smothered in wildfire smog throughout the months of July and August 2010. Visibility was often reduced to 50 m (164 feet), and many famous landmarks, such as St Basil's Cathedral (pictured above), were reduced to mere outlines. The smog led to an increased concentration of airbourne pollutants, six times the normal level of carbon monoxide and twice the normal level of fine particles. Moscow's daytime temperatures reached 38.2°C (100 °F), and over 11 000 city residents died as a direct result of the heatwave.

SHRINKING LAKES, DRYING RIVERS –

reduction in the size of lakes and river flow rates due to climate change or the extraction of water for agriculture or industry

Dry river bed in the Dead Vlei, Namibia

SHRINKING LAKES, DRYING RIVERS

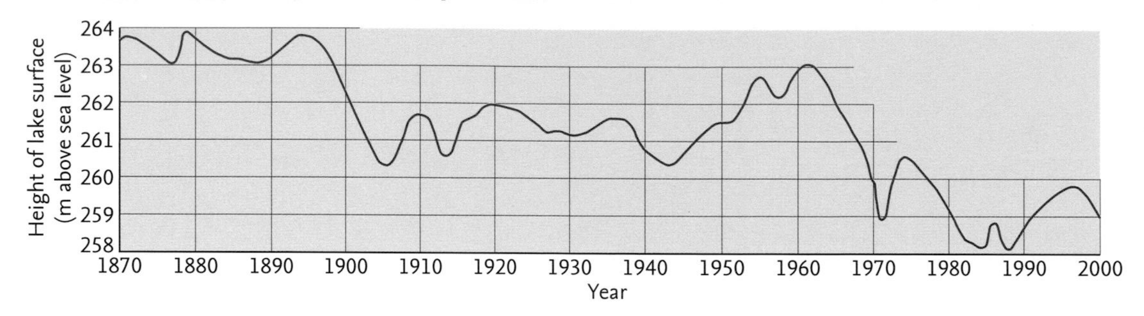

Humans have always based settlements near lakes and rivers. These water bodies provide fresh water, fisheries, transportation and irrigation for crops. But in recent decades, as urban and agricultural demands for water have grown, rivers and lakes across the world have been under assault.

The two rivers which feed into the Aral Sea in Central Asia were dammed in the 1960s to provide water for cotton plantations, devastating the whole area. What was once the fourth-largest inland lake in the world has now retreated into three sections, leaving huge areas of dusty lakebed and old fishing ships stranded on dry land. Only the northern section is thought to be salvageable – the southern Aral Sea will soon dry up for ever.

A similar fate is befalling Lake Chad in the Sahel region of Africa. This lake is now down to 5 per cent of its 1960s size, thanks to a combination of river diversions and perennial drought. In China's Yangtze river valley, more than 800 lakes have disappeared entirely in the last half century. All told, more than half the world's five million lakes are endangered.

However, it is not all bad news. In Mexico, the country's largest body of water, Lake Chapala, had been in long-term decline from the late 1970s. In 1976 the volume of water held in the lake was 8 968 million cubic m (11 730 million cubic yards). By 2003 the volume had declined to 1 307 million cubic m (1 710 million cubic yards). Since then however, the water level has increased to 6 562 million cubic m (8 583 million cubic yards) in 2010 and the quality has improved due to water treatment plants along the Lerma river which flows into the lake.

As well as providing fisheries and fresh water, lakes also provide flood protection by storing excess water and releasing it slowly into rivers. They are also vital parts of the ecosystem, supporting wetland and migrating birds, diverse fish species and other aquatic life. A disappearing lake will often also mean a disappearing animal or plant species, contributing to a global reduction in biodiversity.

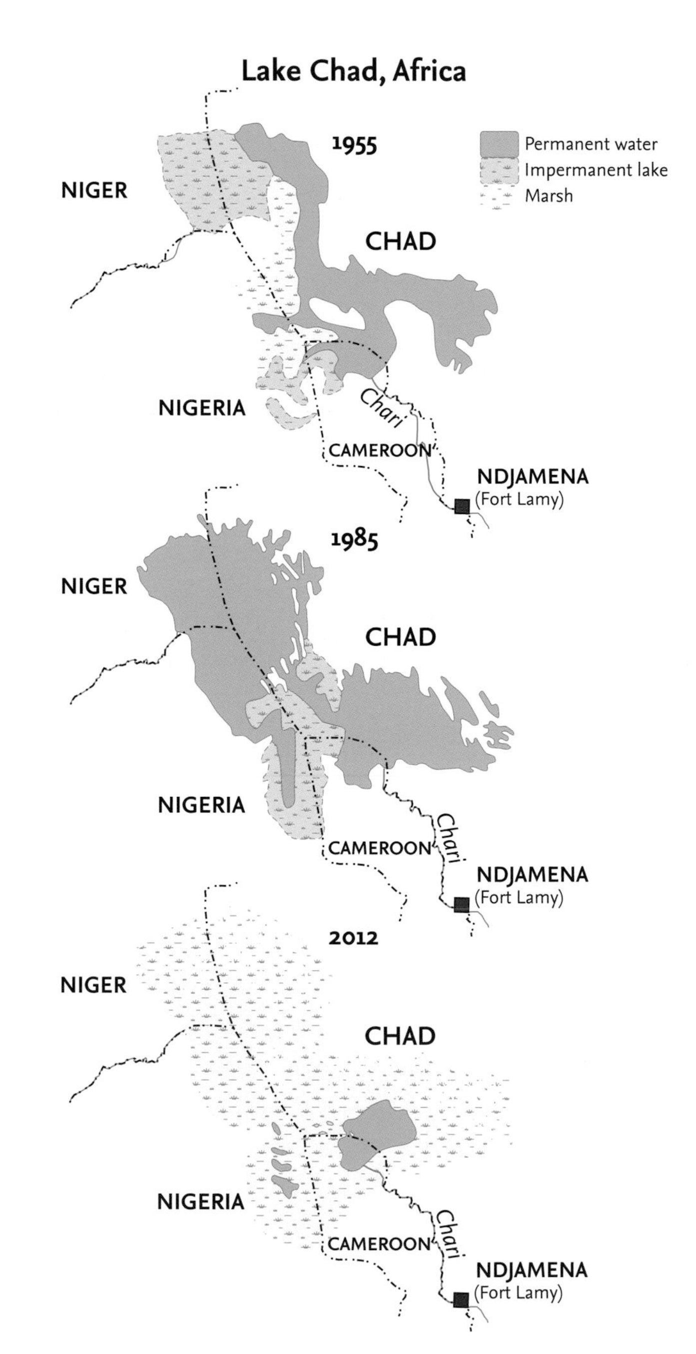

Rainfall variation in the Sahel 1900–2011

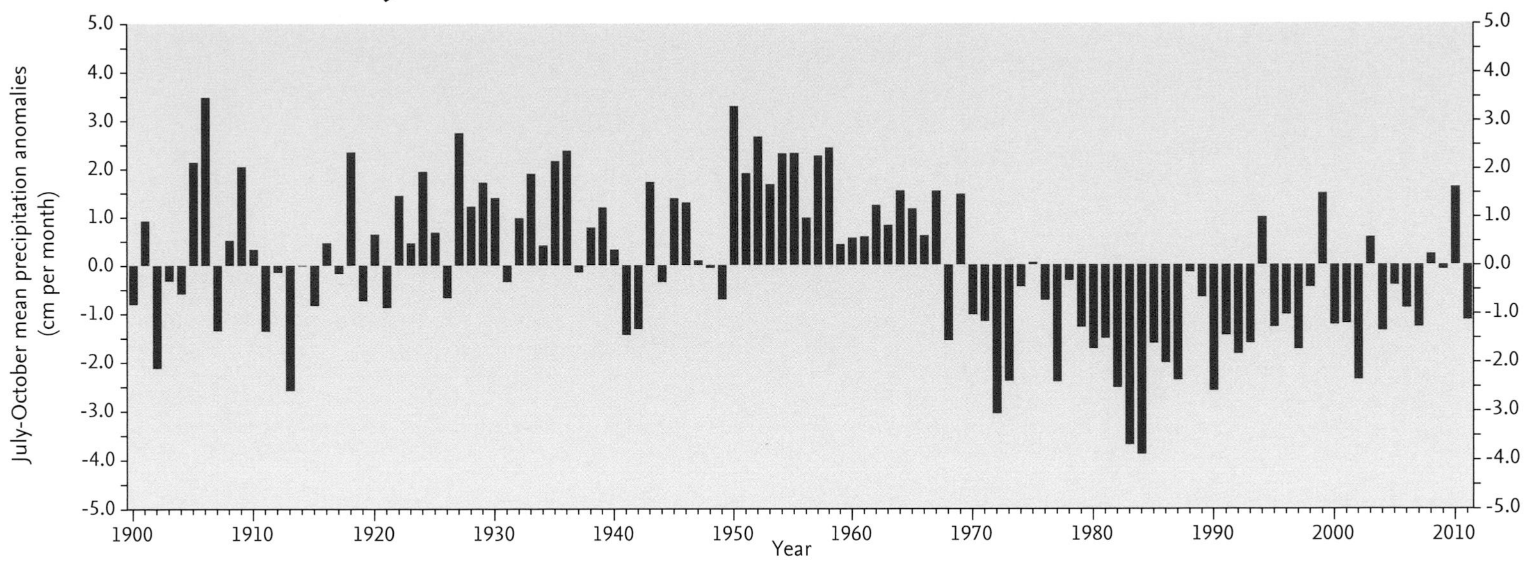

Water supply and use in the Colorado River basin 1923–2006

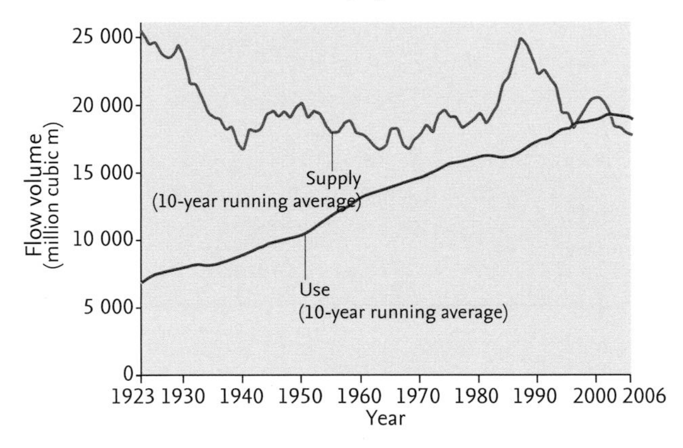

There has been a steady increase in water consumption in the basin since the 1920s but in recent years water use has begun to outstrip water supply. Shortages will occur if this trend continues.

Many of the world's largest rivers now see their flows controlled by people rather than nature. In the United States, the Colorado and Rio Grande river systems have big dams in their catchments which channel water for irrigation and to big desert cities such as Las Vegas. As a result, no water reaches the sea from either river for much of the year. In China, the Yellow River also runs dry most of the time in its lower reaches.

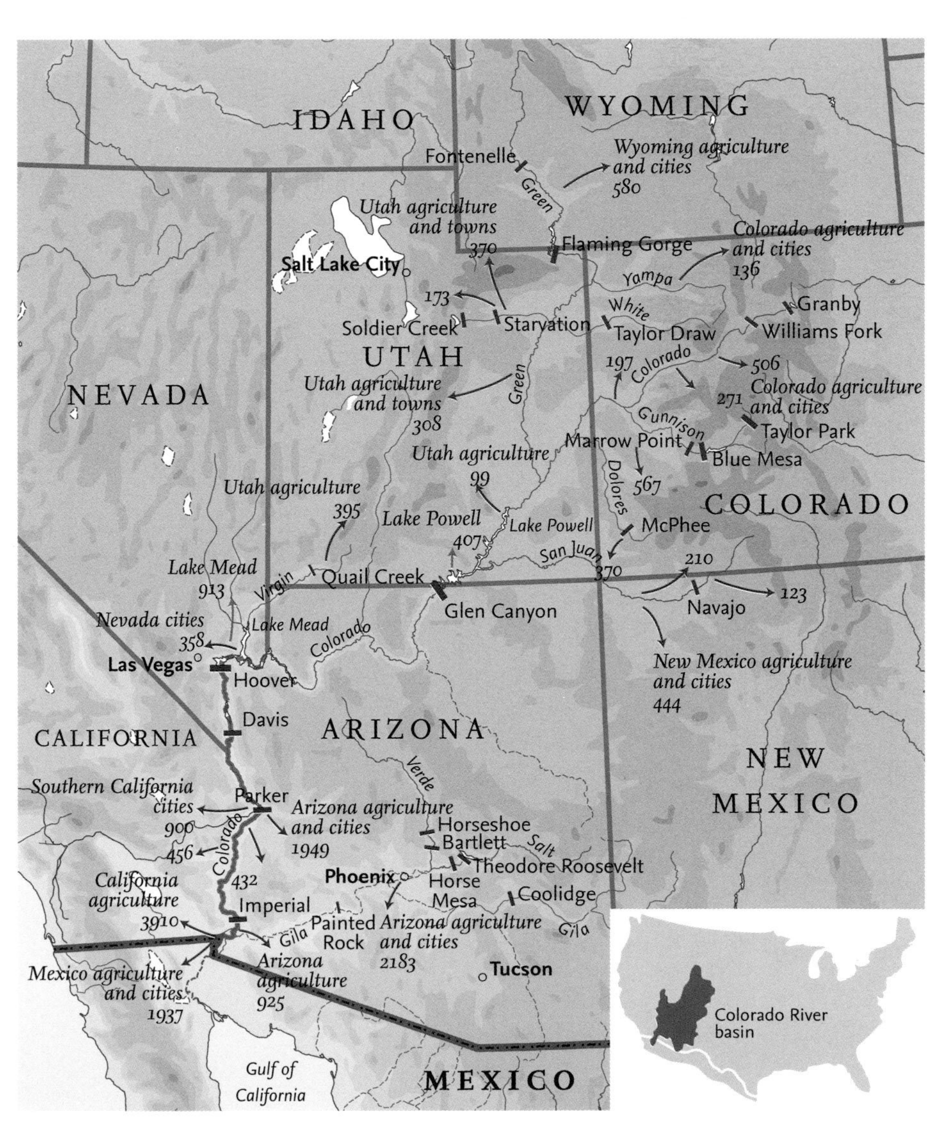

Dam storage in million cubic m
- > 12 000
- 1 200–12 000
- 600–1 200
- < 600

Evaporation and outflow figures are in million cubic m per year (2003–2007)
- → evaporation
- → outflow

1 cubic m = 1.31 cubic yards

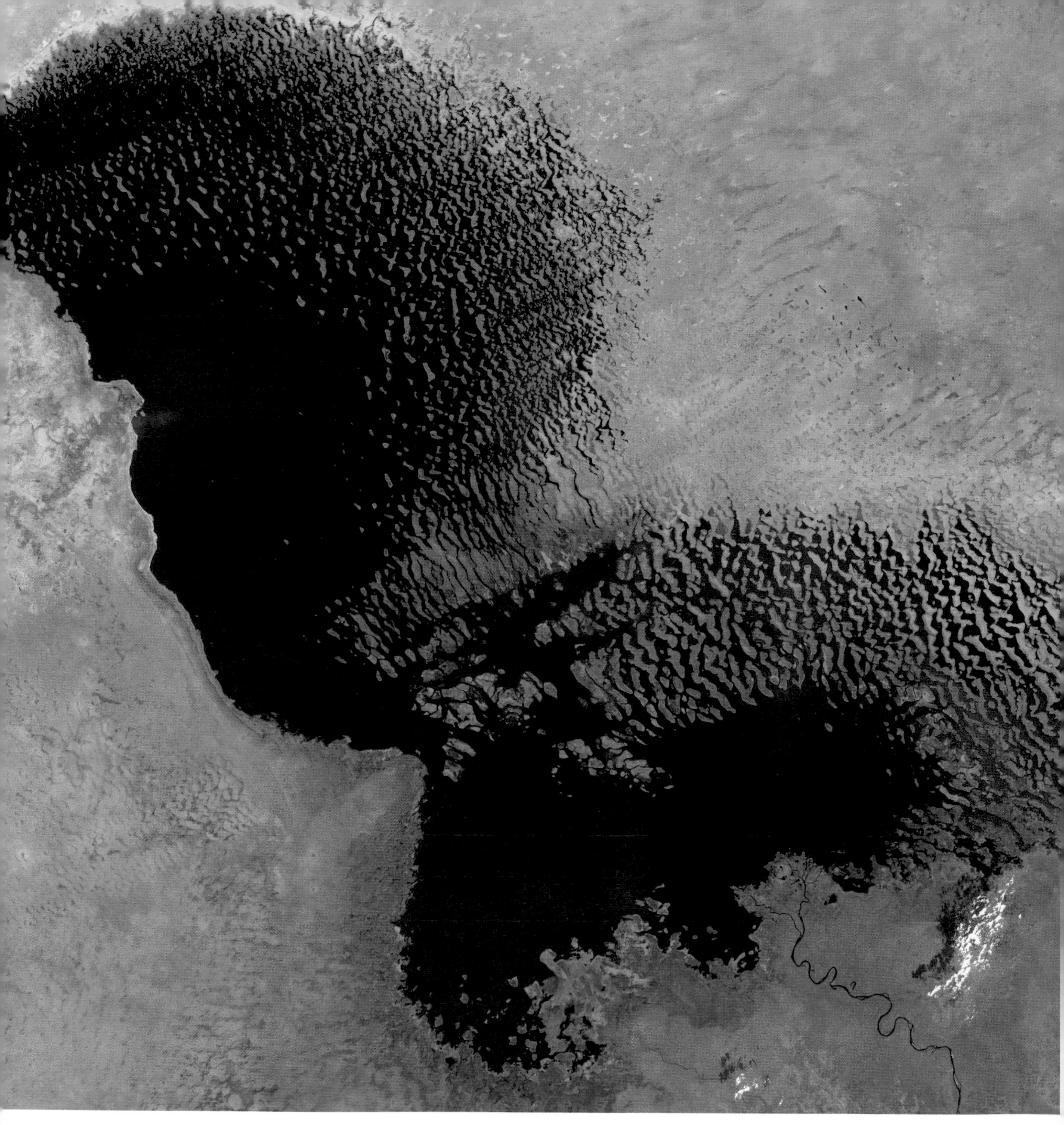

LAKE CHAD, AFRICA 1972

Lake Chad was once one of the largest lakes in Africa, providing water to four countries. But as a result of extensive irrigation projects, the encroaching desert and an increasingly dry climate, it is now a twentieth of its former size. As the lake floor is flat and shallow, the water level fluctuates seasonally with the rainfall. The fifteen years from 1972 to 1987 saw the most dramatic change in the lake, as illustrated in the two main satellite images. This dramatic change was the result of an increase in water being diverted for irrigation.

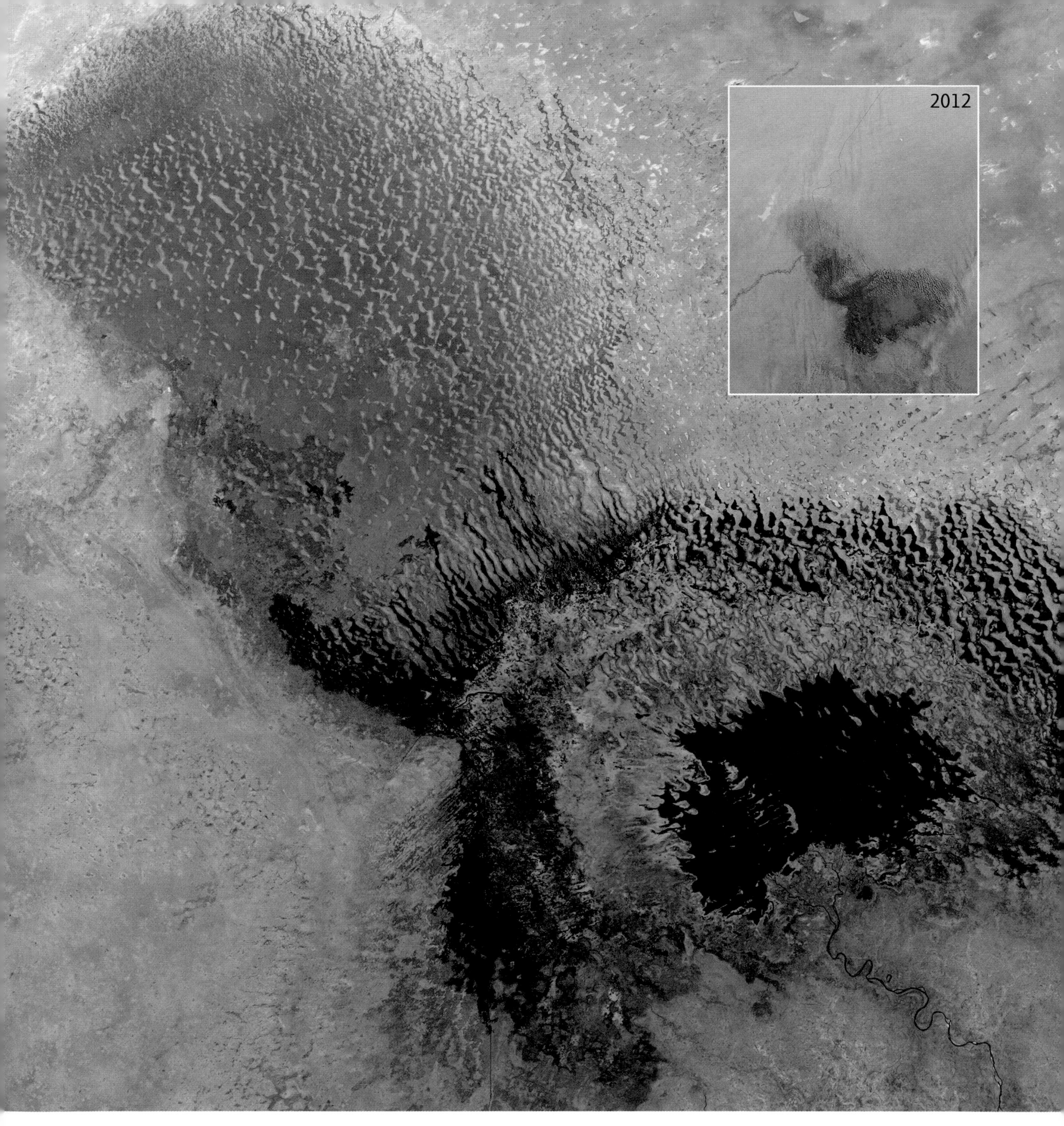

SHRINKING DRAMATICALLY... 1987

However, since the 1980s the water level of the lake has changed very little as a comparison of the 1987 and 2012 (top right inset) images show. Gradually though, with a drying climate, the desert is taking over, as is shown by the ripples of wind-formed sand dunes where the northern half of the lake used to be. Conflict over the water has arisen between fishermen and cattle herders, across borders and within the four countries.

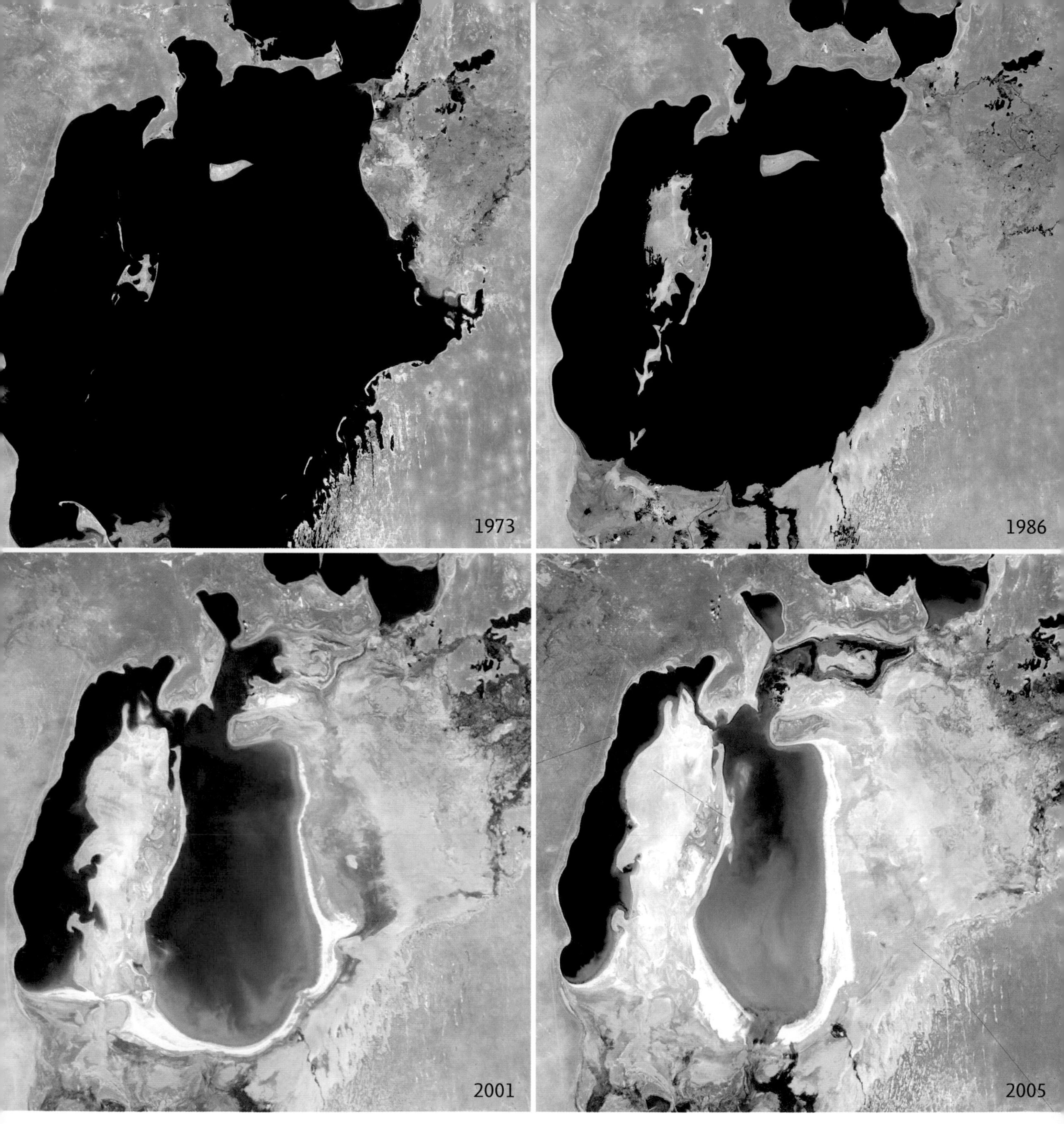

A DYING LAKE, ARAL SEA, CENTRAL ASIA

The Aral Sea was once the world's fourth-largest lake. Today, due to climate change and the diversion of water from its feeder rivers for irrigation for cotton production for example, it is much smaller. Water levels have dropped by 17 m (55 feet). Steps have been taken to preserve the northern part by constructing a dam, but the southern part has been abandoned to its fate.

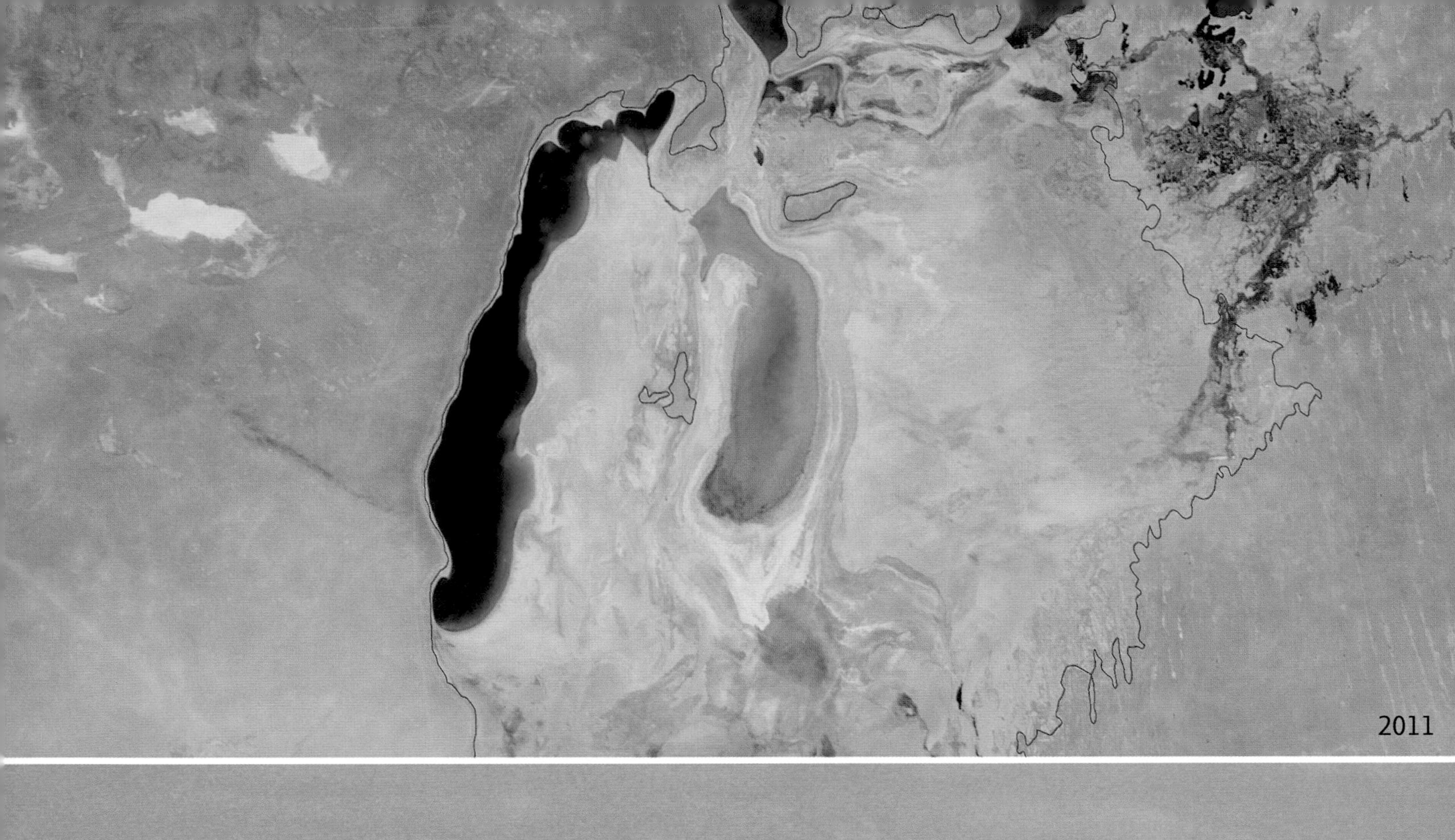

The local fishing industry on the Aral Sea has been devastated by the lake's shrinkage and the local population has developed health problems due to the exposure of chemicals on the dry sea bed. As the lake dries out vast salt plains are forming and dust storms are becoming more frequent.

Abandoned ships litter the former lake bed although the number of rusting hulks is diminishing as they are broken up for scrap in times of economic hardship.

BUCHANAN LAKE, TEXAS, USA 23 OCTOBER 2003...

Buchanan Lake is a man-made reservoir created in the late 1930s and early 1940s by damming the Colorado river. The 3.2-km (2-mile) long dam was built primarily to provide an assured water supply for the lower Colorado river basin, but it also generates hydroelectric power. Buchanan Lake formed behind the dam and is 48 km (30 miles) long, up to 8 km (5 miles) across at its widest point, and has a 200-km (124-mile) long shoreline. The lake is popular for recreational activities such as fishing and boat tours, it also has towns and numerous holiday homes around its shoreline.

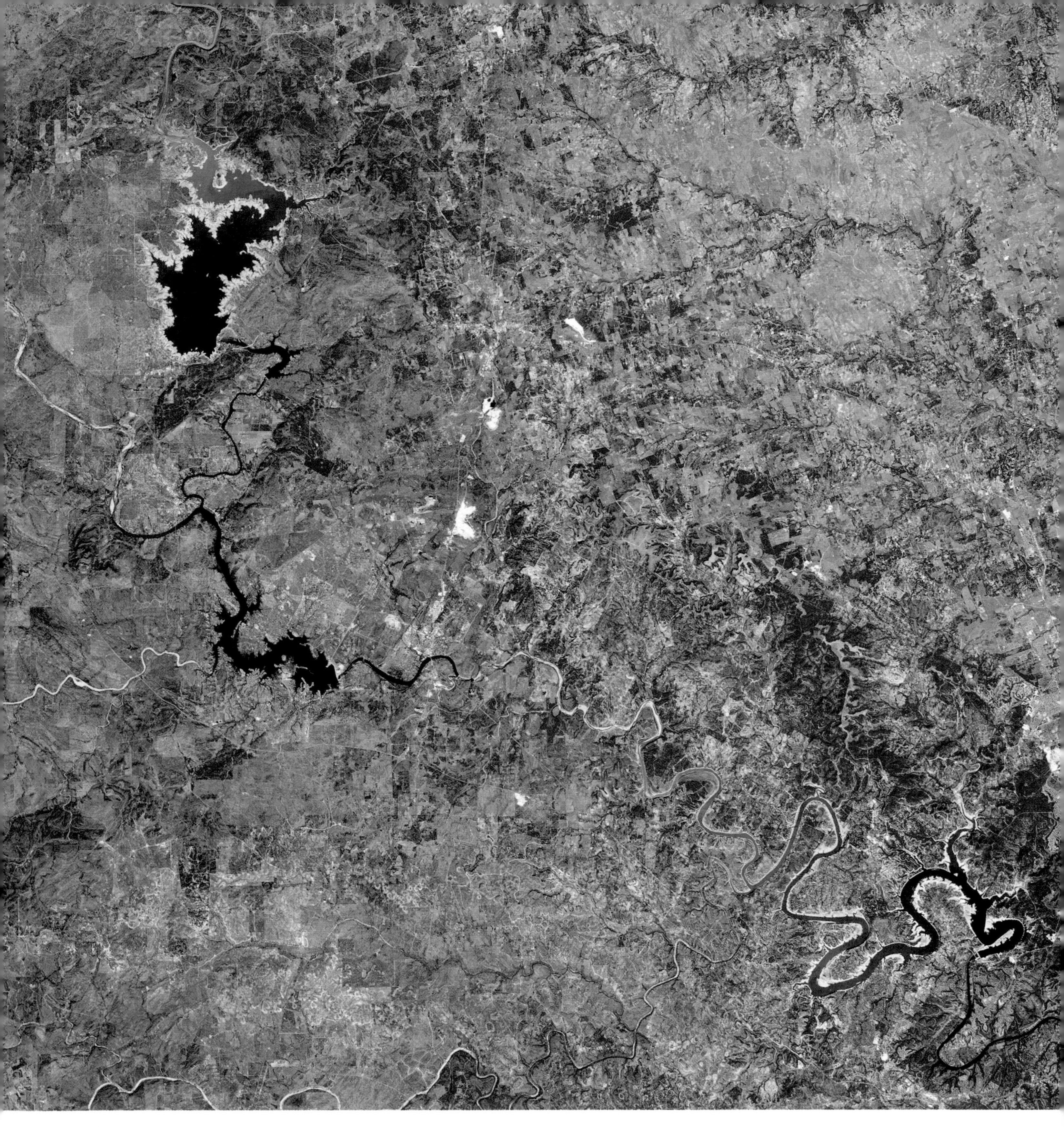

… AND 29 OCTOBER 2011

A year of intense drought in central Texas led to a drop in the lake's water levels and the lake edge retreated almost a mile inland from the stone walls which would normally protect lake-front houses. The shoreline retreated so much that the former town of Bluffton, which was evacuated before being flooded when Buchanan Dam was completed in the 1940s, was exposed once more. Building foundations could be seen along with the remains of the former town's well. By December 2011 the lake's surface level was 7 m (23 feet) below its monthly average.

LAKE URMIA, IRAN 25 AUGUST 1998

Lake Urmia, in northwestern Iran, is one of the world's largest landlocked salt water lakes, and is a designated UNESCO biosphere reserve. Sixty rivers and streams feed the lake and also deliver salts to it. The lake has no outlet; therefore the salts accumulate in the basin, crystallizing along the lake shore when the arid climate causes the lake water to evaporate. Over the last two decades, drought and increased demands for water for agricultural irrigation projects have caused the lake to shrink to less than half of its former size.

SHRINKING DRAMATICALLY… 13 AUGUST 2011

Satellite observations show that Lake Urmia's surface level dropped by an estimated 4 m (13 feet) between 1992 and 2011. As the lake dries up and becomes more saline, its biodiversity declines and it becomes a less-attractive resting place for migratory waterfowl such as flamingos.

There are growing concerns that the increased exposure of the lake bed may lead to windblown dust carrying vast quantities of harmful salts which could endanger human health and prove catastrophic to neighbouring agricultural land.

PARCHED EARTH – POSSIBILITIES FOR A FUTURE

Juliane Zeidler

"I have a dream!" Martin Luther King Jr. described his famous, visionary dream in 1963. He called for racial equality and an end to discrimination. He set out a vision for societal reform, which at that time was unprecedented and difficult to imagine and fathom.

Today, I too have a dream: a dream of a world where we look after the precious natural wonders that are the foundation of our lives. A world in which humanity prospers and continues developing positively, appreciating and nurturing diverse cultures, natures and beliefs against the odds of difficult environmental conditions.

The phenomenon of desertification – a term coined to describe the continued land degradation of previously productive land – has different causes; some of them natural and others man-made. Desertification is not necessarily associated with what we traditionally think of as deserts. Deserts have evolved as a result of long-term geographic and climatic conditions, and often form landscapes and biomes with distinct biodiversity, functional ecosystem services and long-term, rainfall-triggered production patterns, such as plant growth – they are impressive, and highly diverse natural wonders! But they must be cherished for these characteristics and not be exploited beyond their capacities. Desertification often entails the loss of productivity of rangelands, and the consequent suffering of many people through the effects of drought, fire, lack of water and food, and childhood sickness and death.

All of the examples we see in this chapter carry behind them stories of human tragedy. The drying up of the Aral Sea, Lake Chad, the Rio Grande and the Colorado river; the expansion of deserts; the forceful and merciless rage of a veld fire that hurries over the land in a particularly dry year – all affect many, many people and decimate families, belongings and hope for the future.

But there are numerous actions we can take to avoid further man-made catastrophes and help people to prepare for inevitable natural disaster. Humans are innovative creatures – where there is a will there often is a way. While numerous natural disasters can possibly not easily be circumvented or reversed, because either they are a fact of our Earth's nature and activity, or because humankind has already pushed its luck a bit too far, there are many positive actions that can help us protect the fragile Earth as the foundation of our own being:

- We can encourage industries to become more environmentally conscious, and ensure that every business in our global economic world embraces and values sustainability considerations.
- We can lobby our governments and make choices as electorates to ensure that difficult business, social and environmental decisions are made wisely and not just on the basis of a financial bottom line.
- The world governance system, which is so strongly intercepted by multinational business drivers, can be guided towards a compassionate and value-driven performance, mindful of our nature and our limited natural resources.
- We can insist that financial and other investments are made in areas that are at risk and that are often vulnerable, to ensure that people have the option to apply sustainable resource management practices.
- We can practise compassion and contribute to making sure that vulnerable peoples are prepared for disaster, could avoid it, or at least have a level of resilience that provides a possibility for a future.

Appealing to the very inner soul of most people, it is possible to achieve change – change in value systems, change in behaviours, change in decision-making. There are many choices that can be made, and for each of them we can devise plans and put them into action. Reading this chapter and looking at the impressive photography, the amazing satellite imagery, the power and force of nature, the unbelievable extent of human action in terms of environmental impact – underlines that each of us can make choices about what we see, what we want to see and what we want to do about it, if anything.

The parched Earth is not a story of defeat – it is a story of many possibilities for the future. I have a dream...!

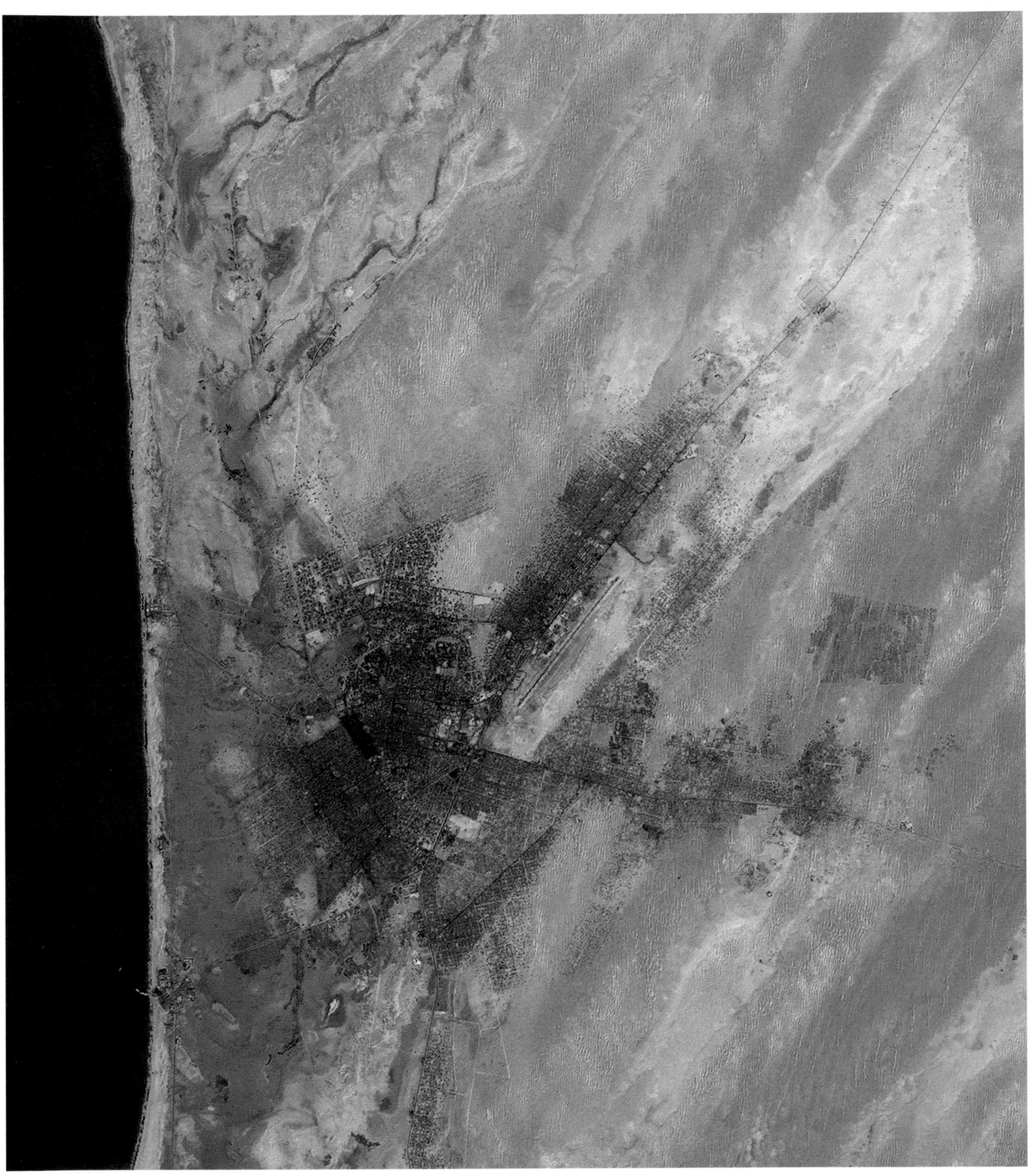

The city of Nouakchott, capital of Mauritania on the west coast of Africa, lies on the edge of the Sahara desert. The dunes are constantly moving and threaten to engulf the city – posing a great challenge to its inhabitants.

WATER'S

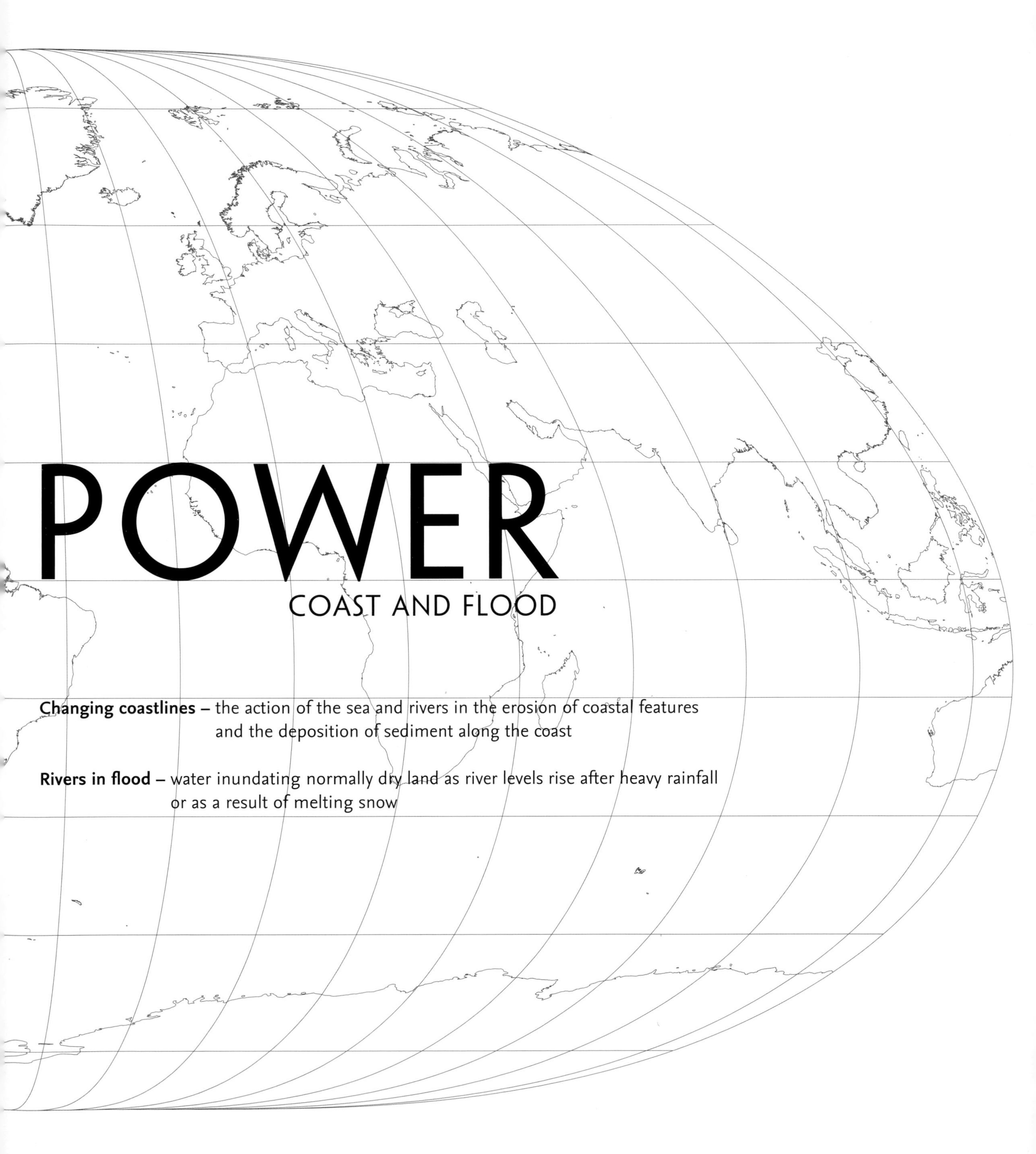

POWER

COAST AND FLOOD

Changing coastlines – the action of the sea and rivers in the erosion of coastal features and the deposition of sediment along the coast

Rivers in flood – water inundating normally dry land as river levels rise after heavy rainfall or as a result of melting snow

CHANGING COASTLINES –

the action of the sea and rivers in the erosion of coastal features and the deposition of sediment along the coast

Pancake Rocks, Punakaiki, New Zealand

CHANGING COASTLINES

The formation of Dungeness, Kent, UK

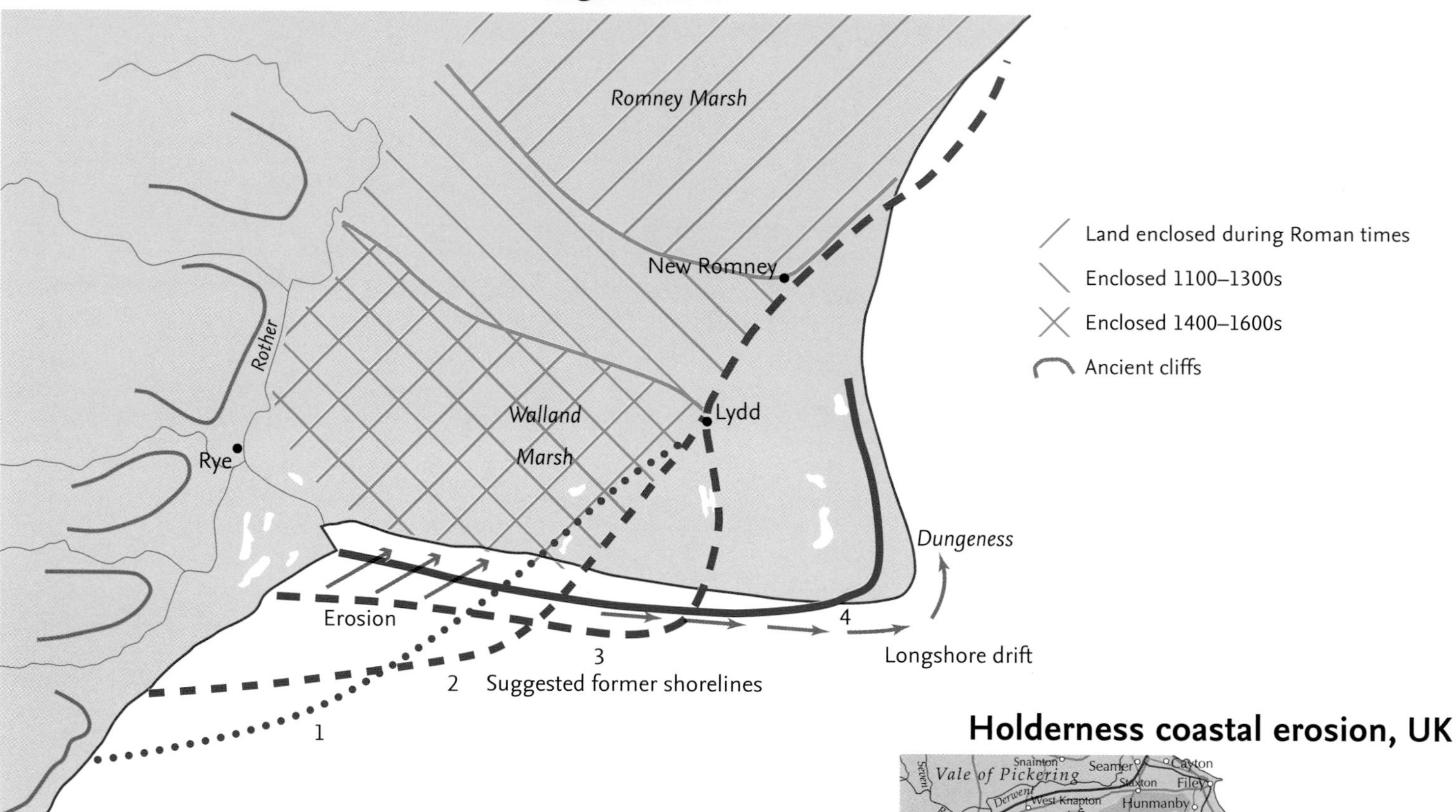

The Earth's coastlines may look permanent to the casual observer, but in reality they are always changing. Land is constantly destroyed by the sea as waves and currents erode beaches and cliffs, but new land is created as sediments are re-deposited by ocean currents elsewhere. Dry land is also formed when rivers create deltas by washing silt out to sea, and – over geological timescales – by tectonic uplift. Volcanoes can create new land very rapidly when they erupt into the sea: lava flows from the eruption of Montagu in the South Sandwich islands, near Antarctica, created 20 hectares (50 acres) of new British Overseas Territory at the end of 2005.

Holderness coastal erosion, UK

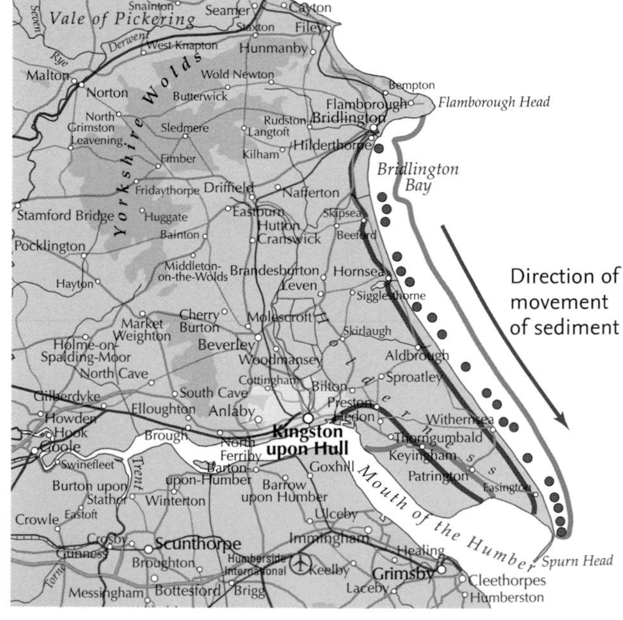

Holderness coastline at risk – one of the fastest eroding coastlines in Europe

Coastal distance lost in 2000 years	5 km (3 miles)
Average rate of erosion	2m (6.5 ft) per year
Largest loss recorded	6 m (19.7 ft) in 2 days (Barmston, October 1967)
Material lost in last 100 years	76 450 000 cubic m (99 992 819 cubic yards)
Deposition at Spurn Head	3% of material eroded
Other deposition areas	Deep water offshore, in Humber Estuary, North Lincolnshire coast
Other losses	Over 30 village sites

Coastal erosion rates

Rock type	Erosion rate mm per year	Erosion rate inches per year
Granite	0.1	0.004
Limestone	0.1–1.0	0.004–0.04
Shale	1	0.04
Chalk	1.0–100	0.04–3.94
Sandstone	1.0–100	0.04–3.94
Glacial till	100–1000	3.94–39.37

Different coastlines erode at different rates according to their constituent materials. Tough rocky cliffs may be relatively impervious to the daily scouring of waves and currents, but mud and clay coastlines erode relatively quickly. The eastern coast of England has been retreating for millennia: the Suffolk town of Dunwich became a 'rotten borough' by the eighteenth century, electing two MPs to Parliament in London even after almost the entire town had been washed into the sea.

Climate change has a direct impact on the coast by driving up global sea levels. Over 70 per cent of the world's sandy shorelines are currently retreating due to sea level rise. Global warming also affects coastal erosion by giving the waves more power as storms get stronger. This is a particular concern for hurricane-prone regions of the world such as the Caribbean and the east coast of the USA.

In the Arctic, coastal erosion is a relatively new problem, because shorelines which used to be in the grip of permafrost and sea ice have now begun to melt and collapse. The Alaskan village of Shishmaref is planning to move off its rapidly-eroding barrier island because melting sea ice has allowed waves to undercut its sandy cliffs (see pages 196–197).

Moving islands – eastern USA

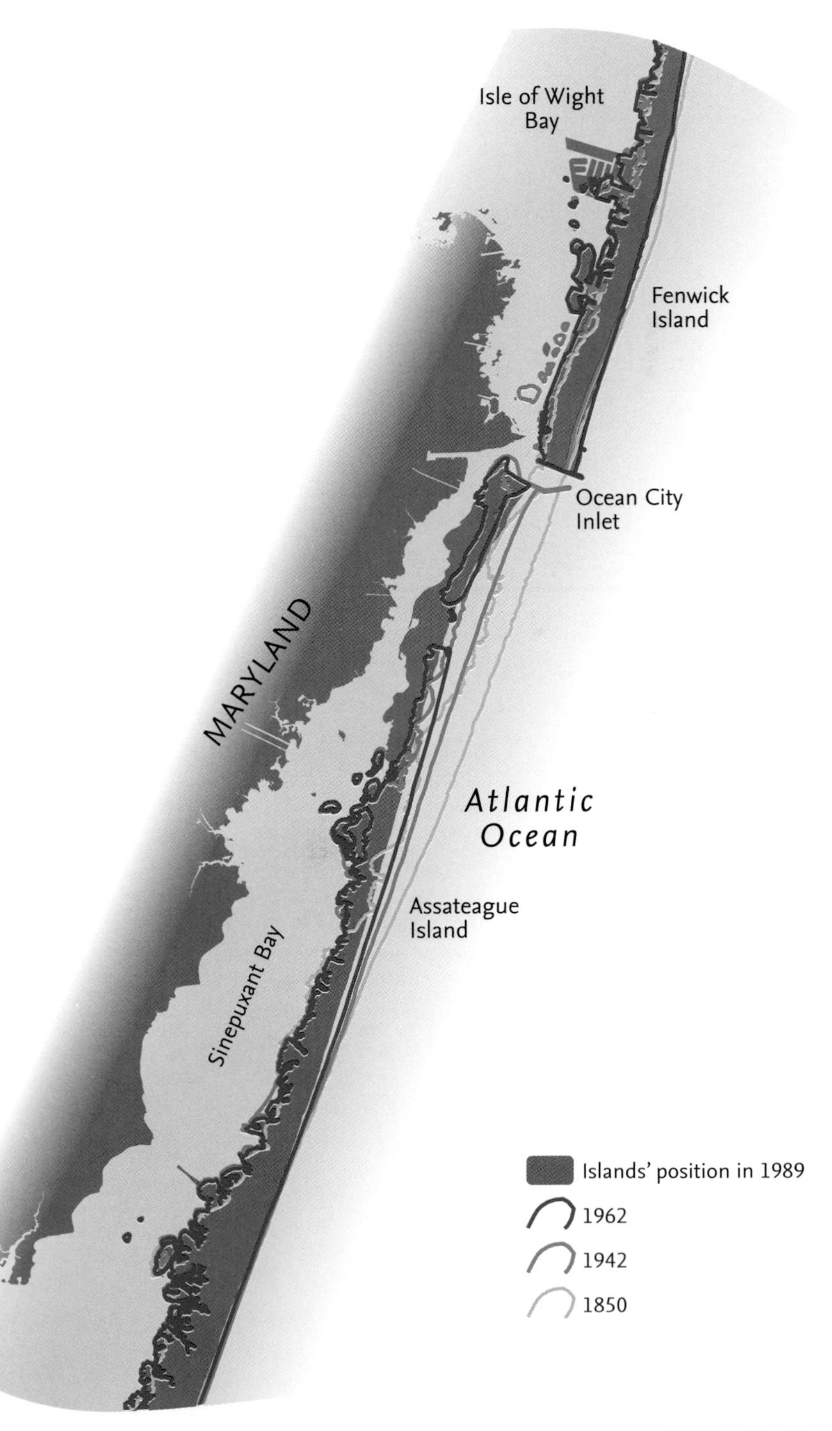

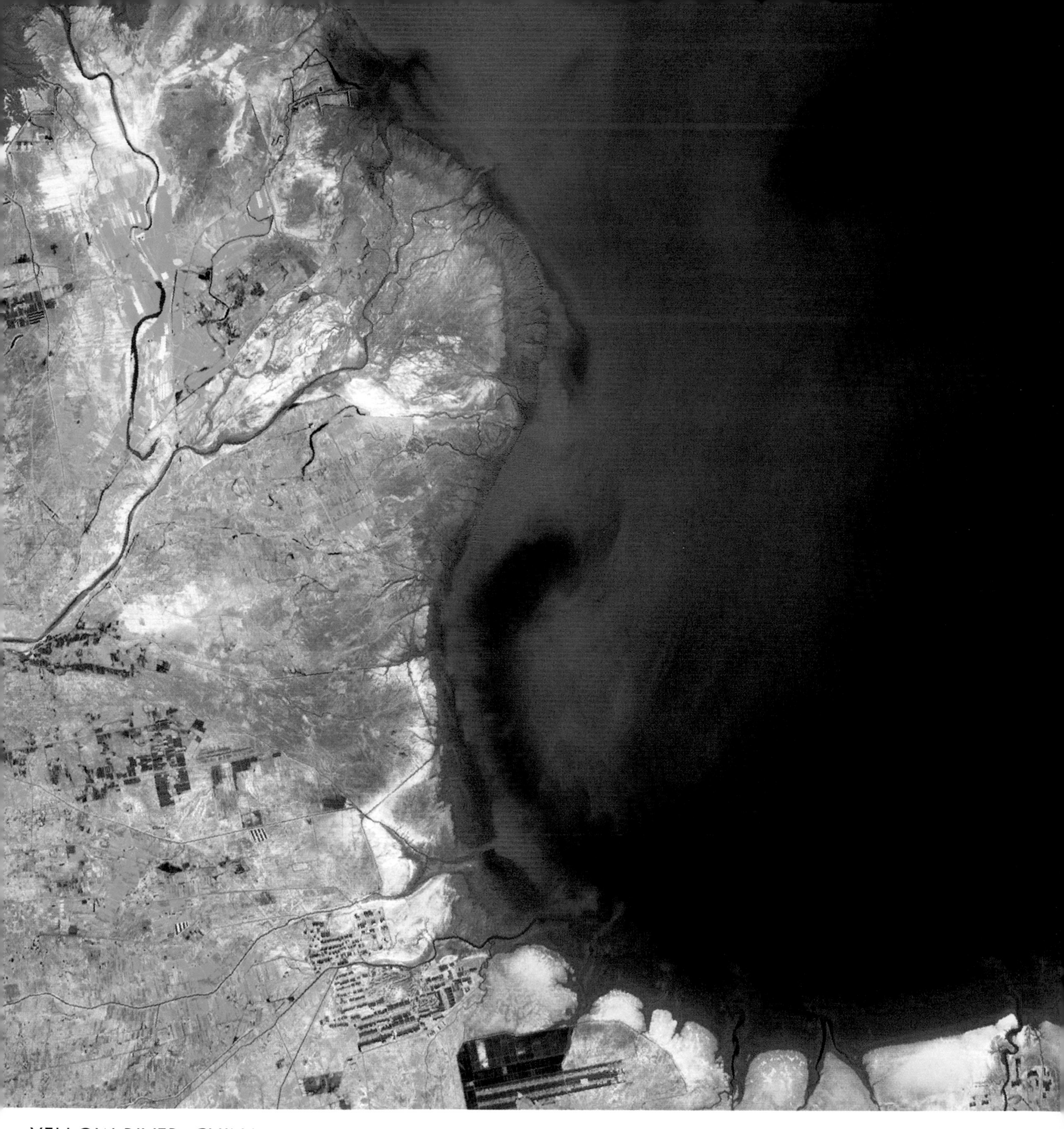

YELLOW RIVER, CHINA 27 MAY 1979

The Yellow River (Huang He) gets its name from the colour of the sediment it carries. This is mainly mica, quartz and feldspar and as the river travels through north central China it crosses an easily eroded loess plateau. Loess is called huang tu, or 'yellow earth' in Chinese, and is a fine-grained deposit which is very susceptible to wind and water erosion. Millions of tonnes were carried down the river each year.

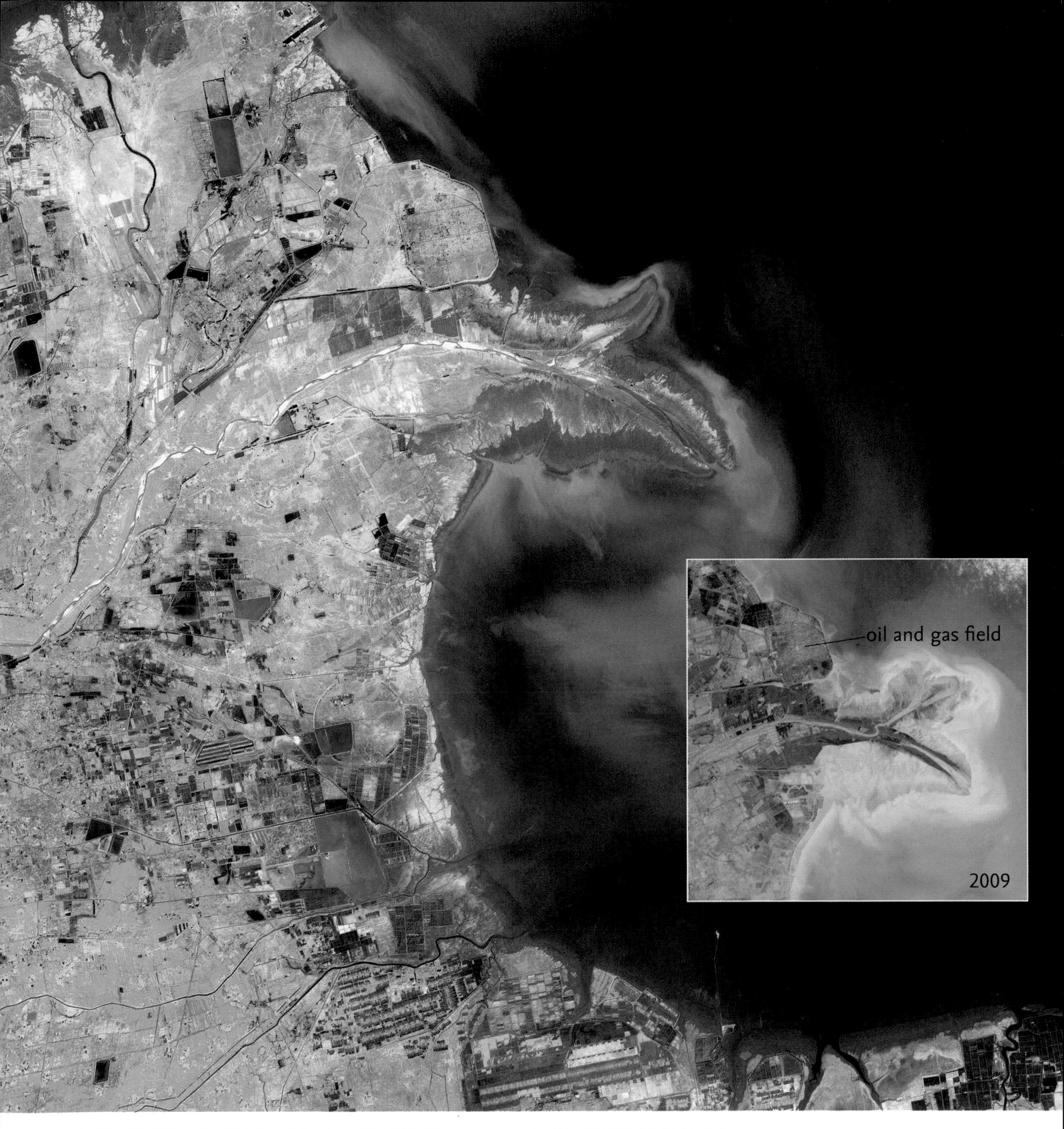

AFTER TWENTY-ONE YEARS OF CHANGE... 2001 AND THIRTY YEARS 2009

On reaching the coast the sediments drop out of the current and are deposited into the river delta. The build-up of sediment, combined with the tidal energy of the sea, has altered the shape of the delta over the years. Since the mid-nineteenth century engineers have tried to control the river by building levees and sea walls to try and protect coastal development. Protection of two local oil and gas fields is an engineering priority. The river has been known to run dry due to a combination of low rainfall and extensive urban development upstream, but the delta continues to change.

THE TWELVE APOSTLES, VICTORIA, AUSTRALIA 9:17 AM 3 JULY 2005

The Twelve Apostles are famous Australian coastal landmarks, which were created by the sea gradually eroding soft limestone cliffs. Cliffs are eroded into headlands, in which caves then form. When caves break through a headland a sea arch is formed, then when an arch collapses a sea stack is left. Erosion at the base of the stacks is around 2 cm (less than 1 inch) each year.

ONE MINUTE LATER... 9:18 AM 3 JULY 2005

The Twelve Apostles are now only eight in number. These two images were taken less than one minute apart on a Sunday morning, when one of the 50 m (164 ft) high stacks gave way to the power of the sea and collapsed into a pile of rubble, much to the shock of onlookers.

CAPE HATTERAS, NORTH CAROLINA, USA 1999 AND 2011

Cape Hatteras is part of a barrier island system on the coast of North Carolina. As a result of rising sea levels, strong storms and warming of the air and ocean, these islands are migrating westward. Between 1870 and 1919 the coast in this area migrated westward by 365 m (1200 feet) and since then there has been a battle to prevent further erosion, including building breakwaters and importing sand. By 2011 we can see how much the coastline of this important tourist area has moved in such a short time.

HAPPISBURGH, NORFOLK, UNITED KINGDOM 2001 AND 2012

The 1950s wooden sea defences at Happisburgh have failed and large chunks of cliff regularly fall into the sea. This part of the Norfolk coastline is soft and vulnerable to coastal storms. The neighbouring settlement of Whimpwell has long since succumbed to coastal errosion. As seen in this pair of photographs, several buildings have vanished over a period of eleven years with homes, businesses, roads and the lifeboat station all at risk. The future of this village is still uncertain, although some protection measures have been undertaken recently.

RIVERS IN FLOOD –

water inundating normally dry land as river levels rise after heavy rainfall or as a result of melting snow

River Trent, Midlands, UK

RIVERS IN FLOOD

River floods can be hugely destructive to life and property, but they play an essential role too. In Bangladesh, images of families forced from their homes by flooding are familiar, but the rising waters are a vital part of an annual cycle – bringing fertility to the fields and replenishing groundwater supplies. Until the building of the Aswan Dam in Egypt, yearly floods brought nutrients to the agricultural land along the river bank. Since the dam began operating in 1964, farmers have had to resort to artificial fertilizers, and parts of the Nile Delta have begun to sink because of the reduction in sediment flowing downstream.

Human interference with rivers makes it very difficult to give an accurate picture of how floods are changing in frequency and magnitude. Recent decades have seen big increases in flood damage, but much of this could be the result of bigger cities and more built-up areas expanding into river flood plains. The construction of levees can help reduce the damage, but if these dykes are breached they can actually work in reverse, preventing water flowing back into the river as it subsides, and making the flooding worse.

Total annual precipitation

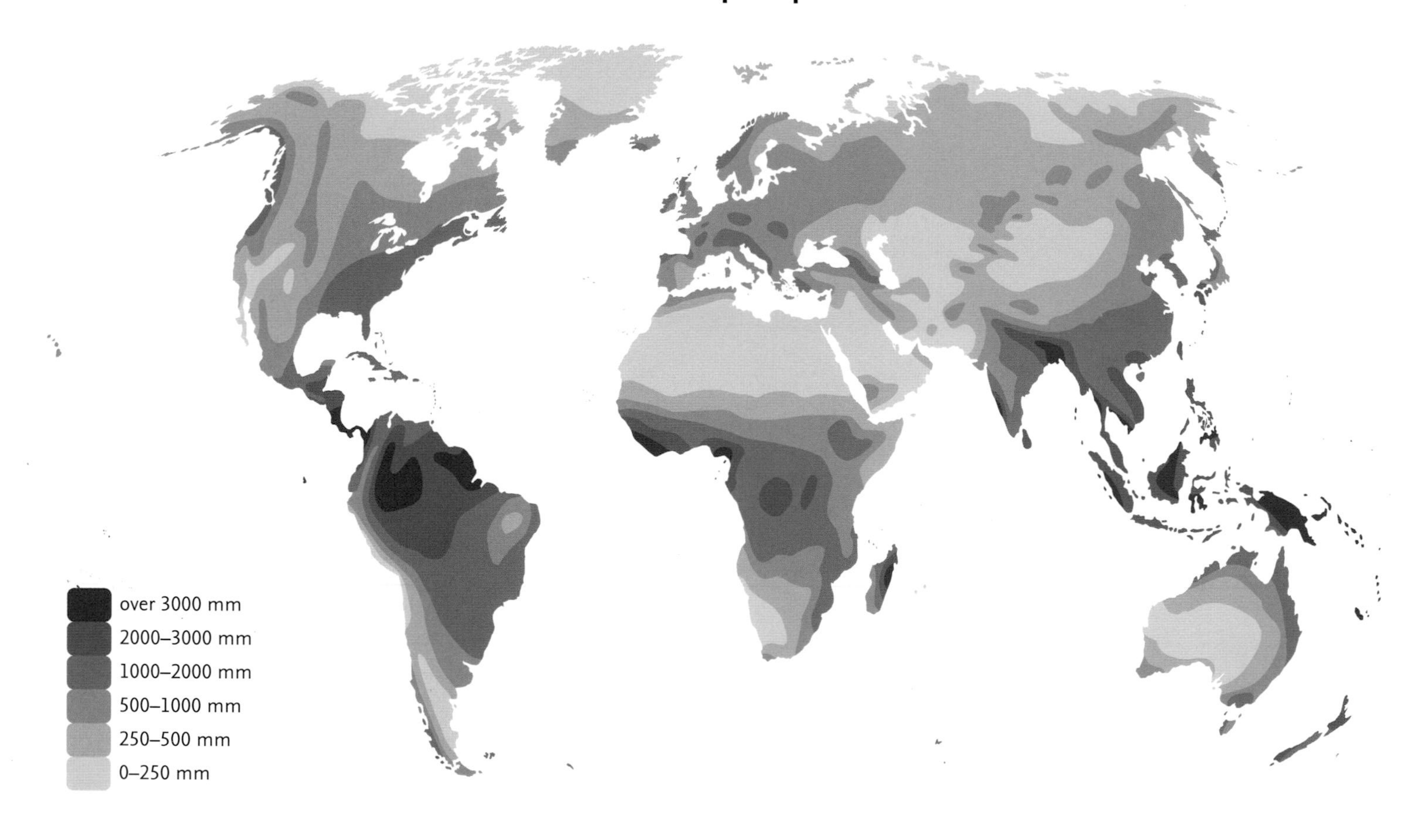

Flooding has a clear relationship with precipitation intensity. Heavier rain and snowfall are undoubtedly linked to a changing climate because a warmer atmosphere can hold more water vapour. Global warming is already making rainfall heavier across the world, from the United States to Japan. In the UK the last thirty years have seen a doubling of rainfall events in the heaviest category, contributing to the devastating floods in October/November 2000. This trend is predicted to continue in the future, with more intense summer monsoons bringing worse flooding to Pakistan, Bangladesh, India and southern China, and heavier winter rainfall affecting higher latitude areas in Europe and North America.

Significant precipitation events 2011

1. Hurricane Irene hits the eastern seaboard of the USA in August leaving flooding, damage to property of US $18.7 billion and 49 deaths.
2. Flooding from storms in June leaves 20 people dead in Haiti, with fears that flood waters will add to the death toll from cholera.
3. After a week of heavy rainfall in October in Central America, 81 people die in floods and landslides.
4. Very heavy rainfall in Rio de Janeiro state, Brazil in January causes landslides which kill more than 900 people.
5. A low pressure area over Europe in November causes severe flooding in France and Italy.
6. In August heavy rains in the Bulambuli district of eastern Uganda cause landslides killing 40 people.
7. Southern Africa experiences unusually high rainfall in January resulting in flooding with over 100 deaths.
8. In September monsoon rains bring catastrophic flooding to Sindh province in Pakistan, killing 509 people and destroying 665 000 homes.
9. Two days of torrential rainfall in June results in flooding and landslides in central China.
10. In July heavy rainfall in South Korea causes mud slides which crush homes on a hillside in Seoul.
11. There is a record amount of rainfall and massive flooding in western and central Japan in September when Typhoon Talas makes landfall.
12. In Sri Lanka flooding and mud slides from heavy rain in January result in 40 deaths and devastation of farmland.
13. In Thailand the death toll from more than three months of flooding reaches 527 by November and flood waters continue to inundate Bangkok.
14. Weeks of flooding along the Mekong river due to unusually heavy rainfall kills 150 people in Cambodia and southern Vietnam.
15. In December tropical storm Washi hits the Philippines causing flash floods and catastrophic damage on the island of Mindanao.
16. Fifty-five people perish and much of Brisbane is left under water in January after flash floods in Queensland, Australia.

Deadliest floods since 1900

Year	Country	Location	Estimated death toll
1911	China	Yangtze	100 000
1931	China	Yangtze	3 700 000
1933	China	Henan, Hebei, Shandong, Jiangsu	18 000
1935	China	East	142 000
1939	China	Henan	500 000
1949	China	Northeast	57 000
1949	Guatemala	East	40 000
1954	China	Yangtze	30 000
1959	China	North	2 000 000
1960	Bangladesh	Central	10 000
1974	Bangladesh	Dhaka	28 700
1999	Venezuela	North	30 000

Indus river flooding 2010

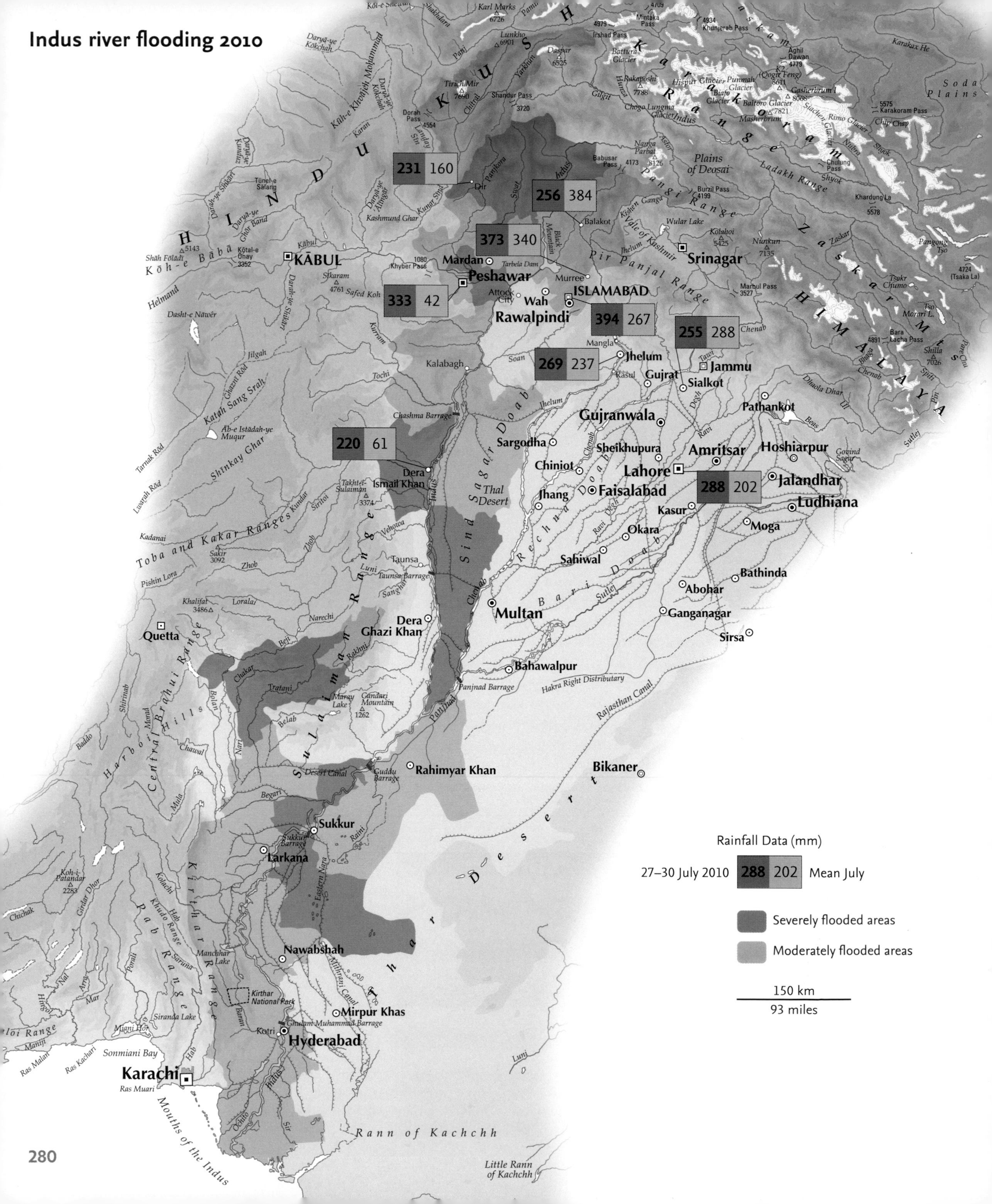

The Indus river basin is home to 100 million people, making it one of the most densely populated river basins in the world. It is also one of the most flood-prone areas, and thousands of people died in floods there earlier in the twentieth-century. Rainfall in the country is highly variable. Northern parts are frequently subjected to heavy showers and snowfall in the winter months, while much of the country is lashed by the Southwest Monsoon in the summer months of June to September. This can lead to significant flooding which tends to be worse in El Niño years, when changing ocean currents in the tropical Pacific spread weather chaos around the globe.

An elaborate system of dykes, canals, dams and floodgates have been built in order to manage the Indus, to reduce downstream flooding, store water to use in times of drought and produce hydro-electricity. Overall, the Indus is much more of a boon than a threat: its waters irrigate farmland helping to keep the country self-sufficient in food. Pakistan is one of the world's leading producers of cotton and a major exporter of rice.

But severe floods in July 2010 illustrated the threat the Indus can still pose. The floods were caused primarily by exceptionally heavy rainfall in July and August in the northern part of the country. The Indus and its tributaries rose rapidly and due to the failure of a dam in Sindh province, part of the river surged down an alternative channel creating a floodwater lake. At least 37 000 sq km (14 285 sq miles) were inundated between July and September 2010. The situation was exacerbated by other factors such as deforestation and over-development. Trees protect headwaters from erosion and forests help to mitigate floods by storing water and releasing it slowly, whereas denuded slopes can see catastrophic flash floods. Once the flood waters had broken through riverbanks, canals and levees, water was then held in by these structures, unable to drain away. Roads, towns and crops were still submerged months after the rains had stopped. Even by late February 2011, floodwaters still lingered on the Indus flood plain.

It is likely that trying to predict future flood patterns in the Indus will become increasingly difficult as monsoon rains will probably become more erratic with climate change.

Flood victims

Year	Deaths	Total affected
2011	509	5 400 755
2010	2 113	20 363 496
2009	102	75 080
2008	83	290 764
2007	526	5 706
2006	400	8 125
2005	636	7 527 043
2004	5	na
2003	266	1 266 243
2002	37	4 010
2001	210	400 179
1999	34	1 043
1998	1 000	200 000
1997	171	851 384
1996	111	1 300 000
1995	1 063	1 855 000
1994	316	840 016
1993	15	261 295
1992	1 446	12 839 868
1991	24	na

Rainfall in northern Pakistan July to August 2009 and 2010

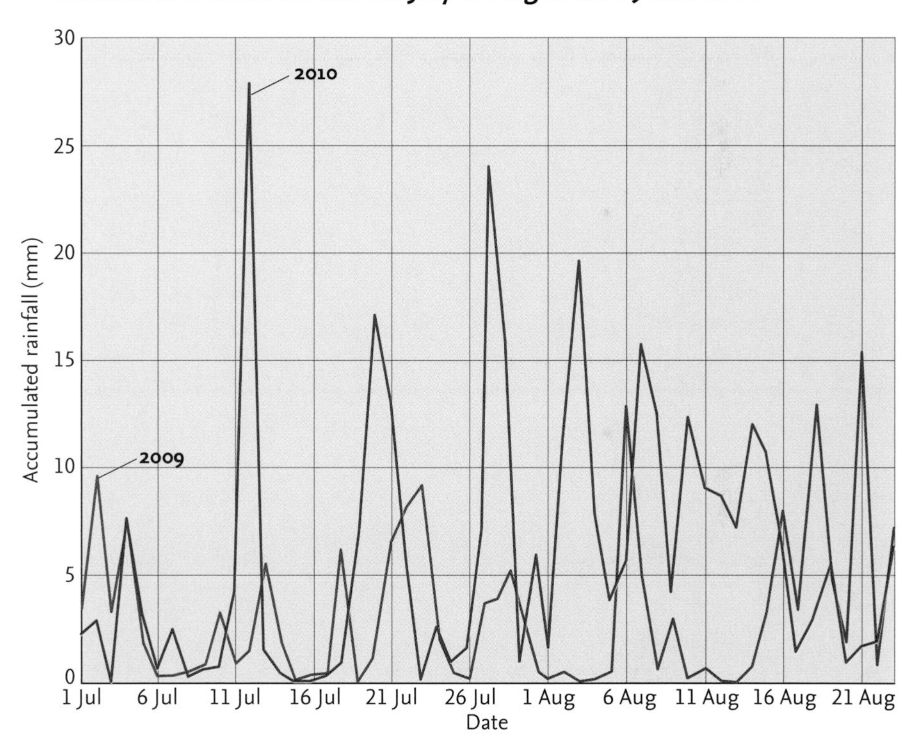

Peak Indus river discharge during monsoon since 1947

	Highest recorded		Second highest recorded		Third highest recorded	
Location	Date	Discharge rate *	Date	Discharge rate *	Date	Discharge rate *
Tarbela	30 Jul 2010	23 559 744	31 Jul 1989	14 441 670	10 Sep 1992	14 158 500
Attock	30 Jul 2010	27 736 501	1958	18 972 390	1966	18 491 001
Kalabagh	30 Jul 2010	26 545 856	2 Aug 1976	24 409 254	10 Sep 1992	24 048 070
Chashma	1 Aug 2010	29 417 766	Aug 1976	22 274 152	29 Aug 1983	19 513 244
Taunsa	2 Aug 2010	27 184 065	1958	22 342 113	7 Aug 1976	19 120 572
Guddu	15 Aug 1976	33 971 112	13 Aug 1986	33 195 792	31 Jul 1988	32 922 845
Sukkur	15 Aug 1986	33 033 875	1976	32 876 037	9 Aug 2010	32 004 439
Kotri	14 Aug 1956	27 759 976	27 Aug 2010	27 322 988	25 Aug 1994	23 400 290

*litres per second 1 cubic foot = 28.3 litres

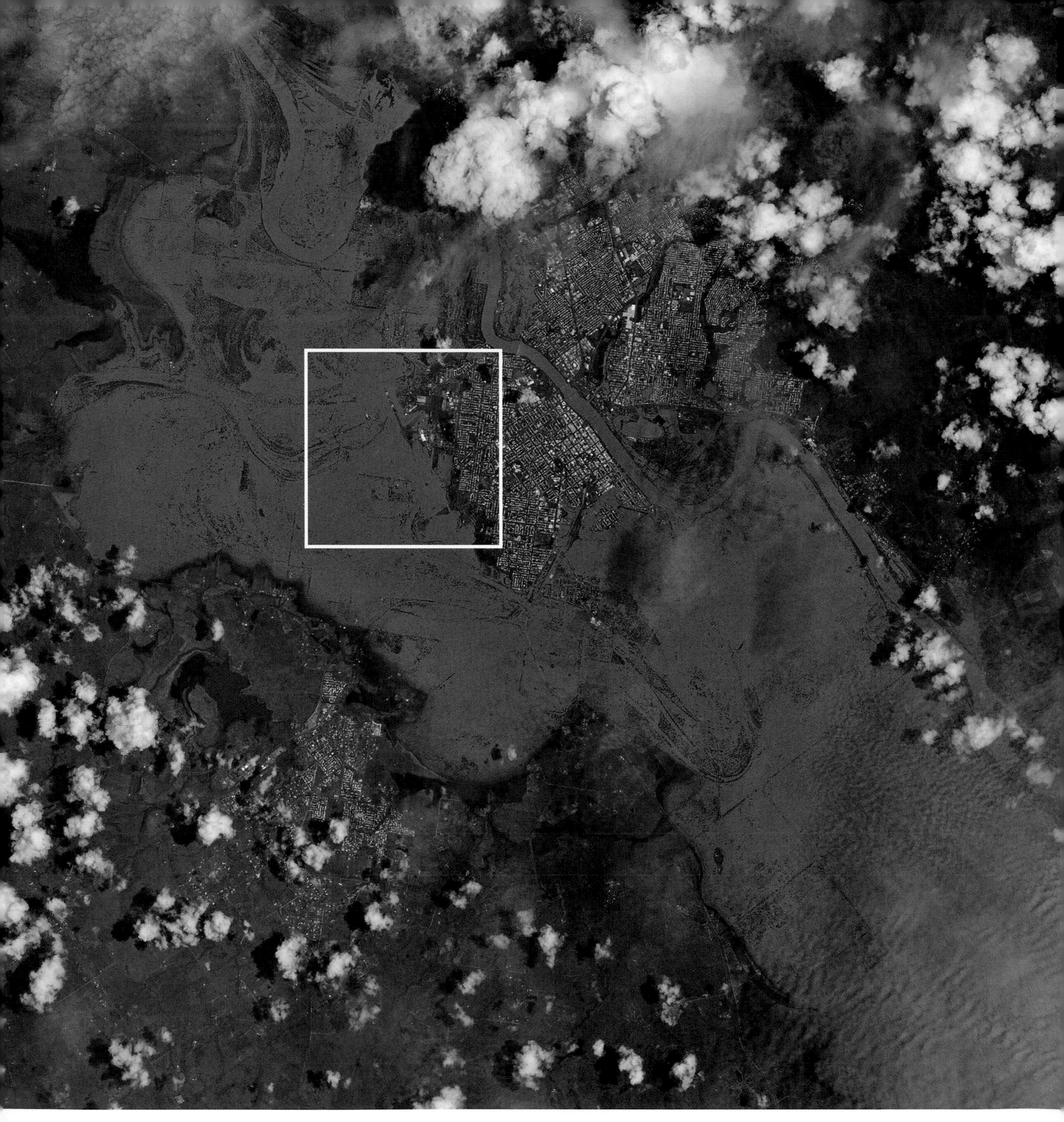

ROCKHAMPTON, QUEENSLAND, AUSTRALIA 9 JANUARY 2011

In the winter of 2010–2011, northeastern Australia was hit by a series of devastating floods. The floods were the result of a combination of factors which intensified the seasonal rains, including a particularly strong La Niña event, and torrential rainfall from tropical cyclone Tasha.

The satellite image above shows the city of Rockhampton a few days after the Fitzroy river, which runs right through the centre of the city, burst its banks. The airport, to the west of the city, was almost completely submerged by the flood waters.

TWO WEEKS LATER... 25 JANUARY 2011

By 25 January 2011 the flood waters began to recede in Rockhampton, leaving behind thick layers of mud, standing water and damaged infrastructure. The Fitzroy river was mostly confined within its banks, and runways were visible once more at Rockhampton airport. The flood damaged thousands of homes and businesses across the city.

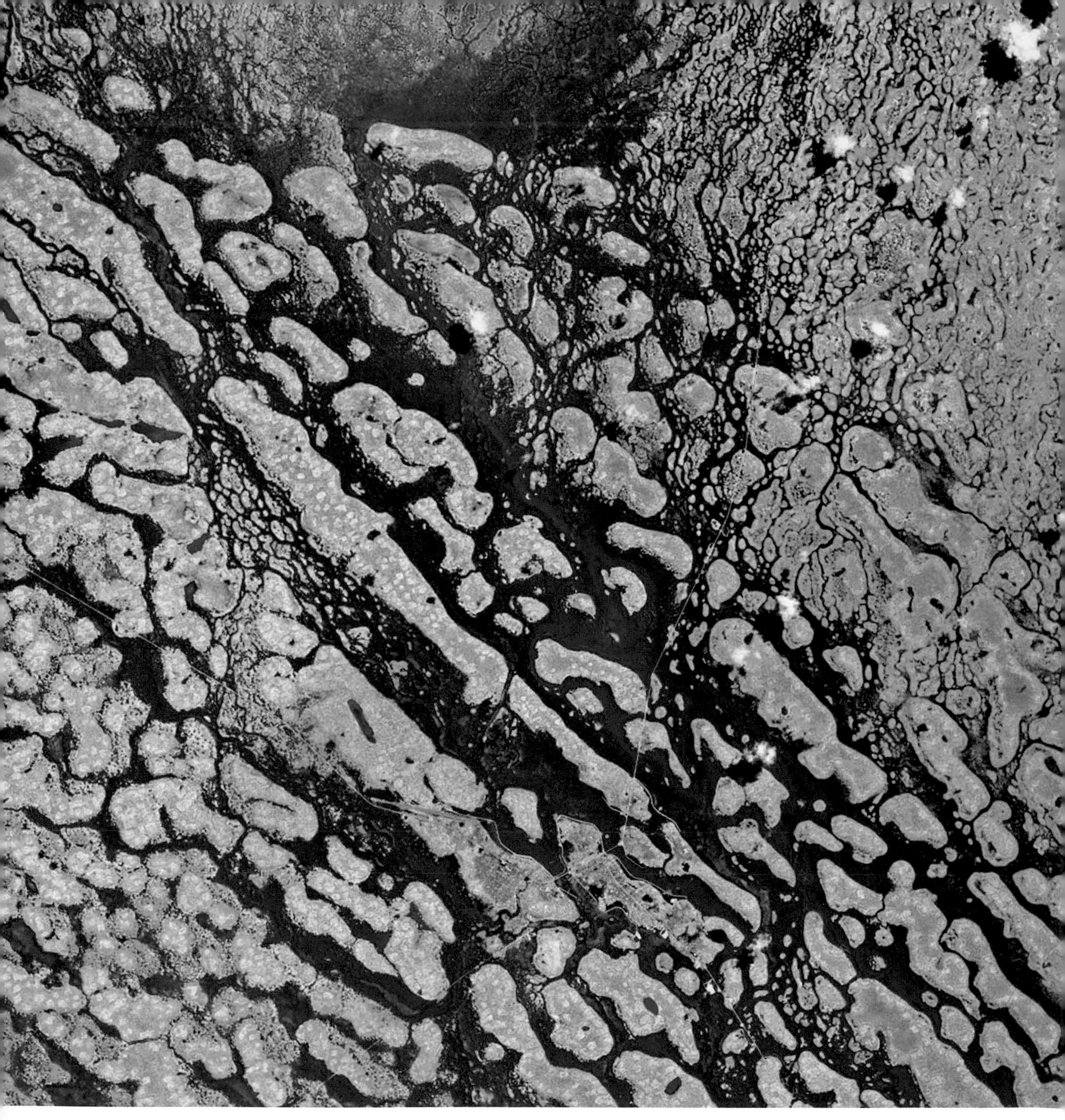

FLOODS AROUND ONDJIVA, ANGOLA 30 MARCH 2011...

Every year seasonal floods affect thousands of people across southern Africa during the rainy season which runs from December to April. In 2011 the flooding was exceptional; above-average rainfall caused severe flooding across Angola by the end of March.

The false-colour satellite image from 30 March 2011, shows the flooding around the city of Ondjiva in southern Angola; water appears bright blue, vegetation is bright green, and bare ground is pink-beige.

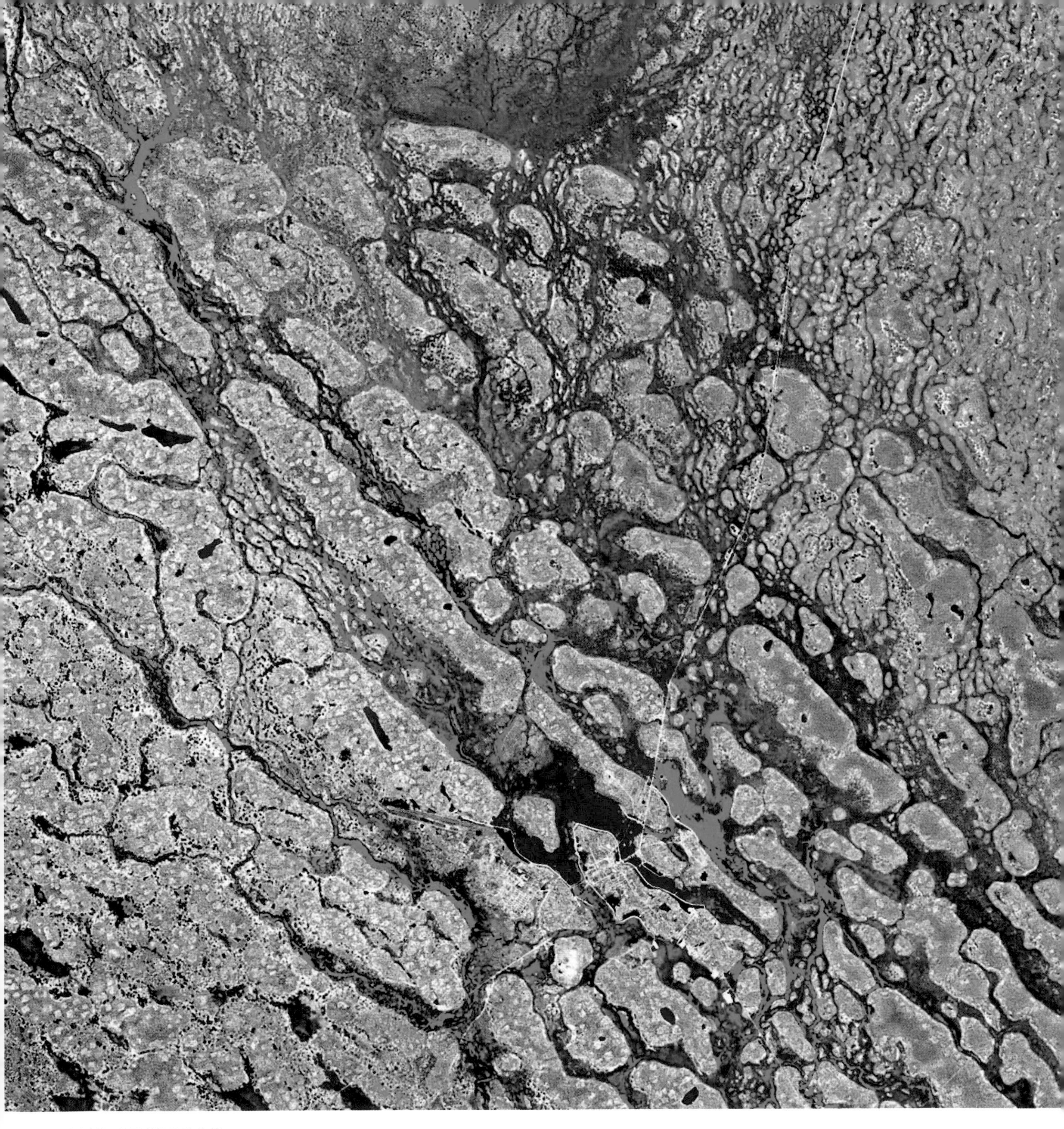

... AND RECEDING 13 MAY 2011

Flood waters eventually receded in May, but not before approximately 260 000 people had been affected, and over 60 000 people had been displaced. The severe floods also caused damage to houses, crops, livestock, and infrastructure with several bridges and roads being swept away.

The satellite image shows the drop in water levels around Ondjiva. Embankments, which appear as white lines on the satellite image, offered some protection to the city from the exceptionally high flood waters.

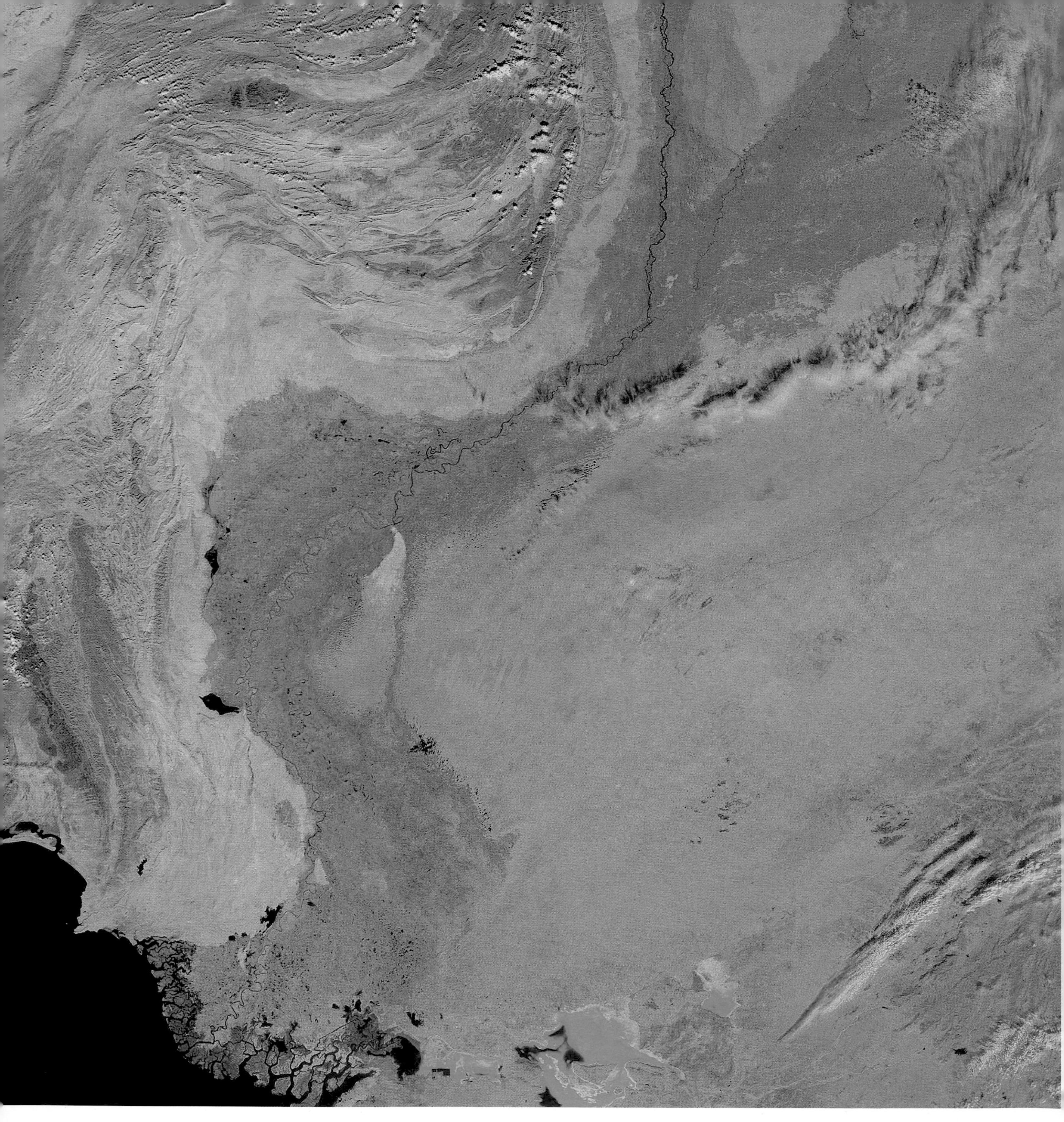

SINDH, PAKISTAN 15 DECEMBER 2009…

Although monsoon rains occur every summer in Pakistan, the summer of 2010 was extraordinary. Devastating floods were the result of several factors including La Niña and an unusual jet stream pattern. The satellite image above shows Sindh in southern Pakistan in the year before the flooding. The meandering Indus river is just visible as a fine blue line running through the fertile green land.

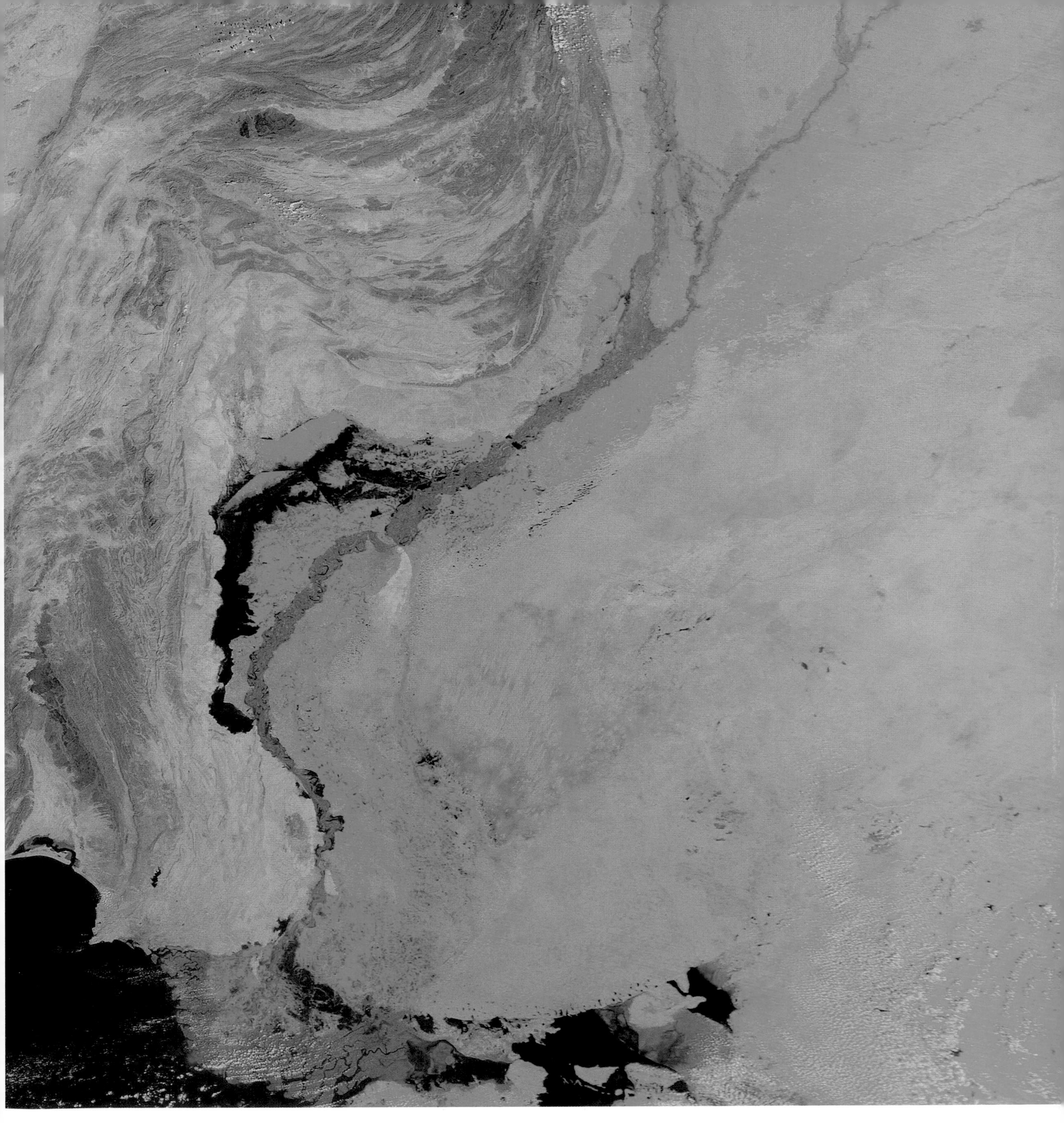

… AND 17 SEPTEMBER 2010

Some of the worst flooding occurred in Sindh, in southern Pakistan; in addition to low-lying land on either side of the Indus river being flooded, a vast new flood-water lake was also created in the province. It formed when flood waters managed to overwhelm a dam just north of the city of Sukkur. The new flood-water lake terminated many miles south in the pre-existing Manchhar Lake. The flood waters retreated very slowly, and in the worst-affected areas whole villages and towns were still completely submerged months later. Infrastructure was obliterated, and crops were ruined.

FLOODING IN SINDH, PAKISTAN 7 DECEMBER 2010

Trees cocooned in spiders' webs were an unusual side-effect of the devastating monsoon floods in Sindh, Pakistan. Millions of spiders climbed up into the trees to avoid the rising flood waters. Five months on, and the flood waters were still present, which led to the rare phenomenon of trees being completely enveloped by spiders' webs. Despite the vast quantities of stagnant water available, the risk of malaria was actually reduced as large numbers of mosquitoes were caught in the huge webs. In July and August 2010, exceptionally heavy summer monsoon rains were responsible

for the worst floods in Pakistan's modern history. The Indus river and its many tributaries burst their banks and flooded the surrounding low-lying areas. In total, nearly two million houses were damaged or destroyed, over 2000 people lost their lives, and approximately twenty million people were affected by the floods. There was extensive damage to infrastructure with villages and roads being submerged for months, bridges being swept away, and thousands of schools and hospitals being damaged.

LIVING WITH FLOODING RIVERS AND CHANGING COASTLINES

LT Patrick Murphy, NOAA

Continuous natural processes link communities along both rivers and coasts. Where water meets land there is a balance between sea-level rise, the ocean's ability to destroy, and new land creation. A dramatic picture of rocky coasts crumbling to the sea, such as the Twelve Apostles in Australia, makes it appear as if beach creation happens overnight. The reality is that nature has been working over a much longer time – years, decades, and even centuries – to maintain the coast.

Natural beach maintenance requires flooding inland rivers or crumbling coastal cliffs to deliver new sediment. As sea level rises, waves move around deposited rock and sand to create landforms such as sandbars, spits, shoals, and barrier islands. When in balance, these formations protect the coast from future storms while also buffering it from further sea level rise.

Along rivers and coasts, communities and governments build structures to prevent flooding and erosion; however, nothing in nature comes free. Efforts in the United Kingdom to protect real estate on cliffs along rocky coasts using sea walls, wire cages, or boulders also prevent natural erosion and the reworking of sediment. On the Mississippi river, levees help prevent flooding and promote safe navigation but also bypass Louisiana's shrinking marshes to deposit sediment offshore. Around the world, dams prevent downstream flooding while maintaining a water supply for farming and drinking. Behind their walls, dams also store millions of tons of sediment that will never make it to malnourished beaches. In many bays, where docks and marinas have protection from the sea, governments dredge channels to allow larger ships to enter port. Like dams trapping sediment, these deep channels trap sand that would normally find its way to the beach. Many barrier islands, including Cape Hatteras on the Outer Banks of North Carolina, are not receiving their historical sand deliveries because routine channel maintenance moves sand out of channels to offshore deposits.

Sea level change and waves reworking the coast are natural processes as old as the ocean which will continue despite man's efforts to the contrary. The balance between sea level, the sea's destructive energy, and land creation rests on a single decision: how much change will we allow?

Governments and communities learned from past failures to tame rising water. Now they are answering the question of change with action. One of the most effective methods is 'managed retreat'. Rather than attempting to hold a line against rising water, this method moves houses, businesses, roads, and utilities inland from river floodplains and eroding coastlines.

Forty years ago, in 1972, Rapid City, South Dakota made a hard decision following a devastating flood which killed more than 200 people and destroyed more than 1 300 homes. The city banned all development on the Rapid Creek floodplain and converted the land along Rapid Creek into a green space now celebrated by residents as the jewel of the downtown area.

In Ventura, California, a managed retreat project is helping to restore the popular surf break Surfer's Point and nearby county fairgrounds. The project's first phase created a bike path and moved a parking lot back sixty-five feet from the ocean. Instead of the parking lot, a cobblestone beach helps trap sediment carried seaward by the nearby Ventura river. A sister project will ensure the river delivers enough sediment to maintain the coastline.

In 1948, the United States government constructed the Matilija Dam seventeen miles up the Ventura river to prevent flooding and store fresh water. Today, the reservoir behind the dam is at 10 per cent of its historic water capacity. The other 90 per cent consists of seventy years worth of sediment – 4.5 million cubic m (160 million cubic feet) – which never made it to Surfer's Point. The coalition managing the retreat project is working with the federal government to remove the now useless dam and return the river to its natural state. Working together instead of fighting the sea, Ventura community groups, landowners, businesses, and the government formulated a plan to restore nature's balance; the results will preserve the aesthetic, recreational and commercial value of a coastal treasure.

Although effective, managed retreat cannot solve every problem. It is one facet of the larger Integrated Coastal Zone Management (ICZM) process formally described at the Earth Summit in Rio de Janeiro, Brazil in 1992. ICZM attempts to unite entities with both common and competing interests in local natural resources. The process recognizes that every issue has different perspectives and no single right answer. ICZM helps both sides start a dialogue and guides them to hard choices. This is still a relatively new idea and many groups are realizing they need to manage their expectations of its ability to fix problems quickly. Nevertheless, this worldwide direction represents a fresh change from the old belief that man can engineer simple, stand-alone defences to hold back a dynamic Earth.

The ribbon of green running through the centre of Rapid City, South Dakota, USA is testimony to restrictions on development of the Rapid Creek floodplain introduced after the devastating floods of 1972.

Top Surfer's Point in Ventura, California in December 1996. The ocean eroded the beach up to and including an adjacent parking area.

Bottom Surfer's Point in Ventura, on 22 May 2012. Phase 1 of the managed retreat project removed the car park and replaced it with a bike path and dunes.

Term	Definition
Aquifer	An underground layer of water-bearing, permeable rock from which groundwater can be extracted.
Avalanche	A fall of snow (or rock) down a mountainside when a build up of snow is released down a slope.
Avalanche crown	The top of the avalanche.
Avalanche flank	The side of the avalanche track.
Avalanche trigger	The mechanism or movement which starts an avalanche.
Biodiversity	A measure of the relative diversity among organisms present in different ecosystems.
Biofuel	Fuel which derives from biomass such as recently living organisms or their by-products.
Boreal	Ecosystems local to sub-arctic and sub-antarctic regions.
Carbon cycle	The exchange of carbon between the biosphere, geosphere, hydrosphere and atmosphere of the Earth.
Catchment area	A drainage basin or region of land whose water drains into a river or into a body of water.
Coriolis effect	A force resulting from the rotation of the Earth, affecting winds, ocean currents and atmospheric conditions.
Crown fire	A wildfire which spreads to the top branches of trees sucking oxygen upwards.
Crust	The outer solid layer of a planet. The Earth's crust is between 5 and 60 km (3 and 37 miles) thick.
Cyclone	The rotation of a volume of air around an area of low atmospheric pressure.
Debris toe	Where an avalanche ends when the flow stops.
Deforestation	The conversion of forested areas to non-forested, by the removal of trees usually to create agricultural land.
Desertification	The degradation of land in arid, semi-arid and dry sub-humid areas from climate change and human activities.
Drought	An extended period where water availability falls below the requirements for a region.
Earth-observing satellite	A satellite equipped with sensors which observe and record information about the Earth's surface.
Earthquake	A sudden and sometimes catastrophic movement of a part of the Earth's crust.
Ecosystem	The smallest level of organization in nature which incorporates both living and non-living factors.
El Niño	Marked warming of the Pacific Ocean which occurs when an unusually warm current of water moves from the western Pacific, temporarily replacing the cold Peru Current along the west coast of South America.
Evapotranspiration	The return of water to the atmosphere by evaporation from the soil and water bodies and by transpiration from plants.
GDP	Gross Domestic Product. The value of all goods and services produced within a country.
Global warming	An increase in the average temperature of the Earth's atmosphere and oceans which has been observed in recent decades.
Greenhouse effect	The process by which an atmosphere warms a planet.
Greenhouse gas	Gaseous components of the atmosphere which contribute to the greenhouse effect. The major natural greenhouse gases are water vapour, carbon dioxide and ozone.

Term	Definition
Ice cap	A dome-shaped ice mass which covers less than 50 000 sq km (19 305 sq miles) of land area.
Irrigation	Artificial suply of water for agriculture in otherwise fertile areas.
Land reclamation	The creation of new land where there was once water.
Landslide	A sudden movement of a mass of soil or rocks down a cliff or slope.
Levee	A natural or artificial embankment or dyke, usually made of earth, which parallels a river course and protects the adjacent area from flooding.
Mantle	The thick shell of rock surrounding the Earth's outer core, lying between 30 and 2900 km (19 and 1802 miles) below the Earth's crust.
Mudslide	Downward movement of a mass of sediment liquified by rain or by melting snow or ice.
Permafrost	Ground which is permanently frozen.
Pivot-point irrigation	A method of crop irrigation where the equipment rotates around a central pivot giving a circular crop pattern.
Polder	A low-lying tract of man-made land, enclosed by embankments or dykes.
Precipitation	Rain, snow and other forms of water falling to the Earth's surface
Pyroclastic flow	Fast moving bodies of hot gas, ash and rock (collectively known as tephra) emitted from certain types of volcanic eruption.
Sahel	The boundary zone in Africa between the Sahara to the north and the more fertile region to the south.
Sea ice	Ice formed when sea water freezes. Fast ice has frozen along coasts, pack ice is floating consolidated sea ice and an ice floe is a floating mass of sea ice.
Sea level	Mean sea level (MSL) is the average height of the sea, with reference to a suitable reference surface.
Seismic wave	A wave which travels through the Earth, often as the result of an earthquake or explosion.
Stauchwall	The downslope fracture surface of an avalanche.
Storm surge	An onshore rush of water associated with a low pressure weather system, typically a tropical cyclone.
Subduction zone	An area on Earth where two tectonic plates meet, move towards one another, with one sliding underneath the other and moving down into the Earth's mantle.
Tectonic plate	A large piece of the Earth's crust.
Tornado	A violently spinning column of air in contact with both the cloud base and the surface of the Earth.
Tsunami	A series of waves generated when a body of water, such as a lake or ocean is rapidly displaced on a massive scale.
Tundra	Treeless zone between the Arctic ice cap and the northern treeline which has a permanently frozen subsoil (permafrost).
Urbanization	The movement of people from rural to urban areas.
Volcano	A mountain formed from volcanic material ejected from a vent in a central crater.
Water table	The level below which the ground is saturated with water. It rises and falls in response to rainfall and the rate of water extraction.
Wildfire	An uncontrolled fire often occurring in rural or forested areas.

Page numbers in **bold** refer to captions, illustrations and photographs.
Page numbers in *italics* refer to information contained in tables.

Contributors

Julia Bucknall
Sector Manager, Water Department, World Bank, known as the World Bank's Water Anchor. Lead author of a flagship publication on water in the Middle East and North Africa region, *Making the Most of Scarcity*, and has experience in water investment projects in Africa, Asia, Europe and Central America.

Professor John Grace
Research scientist and environmental consultant, now Emeritus Professor at the University of Edinburgh, Scotland. He is the author of over 200 science papers, many of them dealing with aspects of global change.

Professor Mark Maslin
Professor of Climatology at University College London, Royal Society Industrial Fellow and Executive Director of Carbon Associates Ltd. He is a leading scientist with expertise in global and regional climatic change, widely published and a regular media presenter.

Professor Bill McGuire
Professor of Geophysical & Climate Hazards at University College London, science writer and broadcaster. His current research focuses on the relationship between climate change and geological hazards. Bill was a member, in 2005, of the UK Government Natural Hazards Working Group and was a contributing author to the 2011 IPCC report on climate change and extreme events.

Lieutenant Patrick Murphy
A member of the United States National Oceanic and Atmospheric Administration Commissioned Officer Corps. Author of several peer-reviewed papers, his assignments include time ashore with NOAA's National Marine Fisheries Service and time underway as a deck officer aboard NOAA's research vessels.

Dr Juliane Zeidler
Managing Director of Integrated Environmental Consultants of Namibia (IECN). Has been working in the fields of environment and development for the past fifteen years and is known internationally for her expertise in biodiversity research, desertification, natural resources management, community development, environmental politics and sustainable development.

Original contributions to the first edition by:
Michael Allaby. Environmental science writer.
Mark Lynas. Broadcast commentator and journalist.
Fred Pearce. Environmental consultant.

General Acknowledgements

We would like to acknowledge the assistance of:

The United Nations Environment Programme (www.na.unep.net) in providing selected images from *One Planet Many People - Atlas of Our Changing Environment*, UNEP, 2005 (www.earthprint.com).
Page 138 Christopher Riches for kind permission to reproduce an image from *Picturesque Hongkong*. Photos by Denis H. Hazell, published by Ye Olde Printerie Ltd, Hongkong, circa 1920.
Pages 216 University of Colorado, CO, USA for image of global mean sea level 1993–2012.
Univ of Colorado 2012_rel2 sealevel.colorado.edu.
Page 149 World Bank for permission to use images from the World Bank collection.

GeoEye www.geoeye.com
European Space Agency www.esa.int
National Snow and Ice Data Center, University of Colorado, Boulder, CO, USA nsidc.org
NASA earthobservatory.nasa.gov
NASA rapidfire.sci.gsfc.nasa.gov
NASA asterweb.jpl.nasa.gov/index.asp
United States Geological Survey www.usgs.gov
World Bank www.worldbank.org

Image credits

12–13 Alamy/Pacific Press Service/Masanori Kobayashi
18 U.S. Coast Guard photo by Petty Officer 2nd Class Sondra-Kay Kneen
19 Satellite image courtesy of GeoEye
19 **top inset** M_Eriksson
19 **bottom inset** United Nations Development Programme
20 Earl and Nazima Kowall/Corbis
21 Ryan Pyle/Corbis
22 */AP/Press Association Images
23 Martin Hunter/Getty Images
24 © Anatolia/Xinhua Press/Corbis
25 Aamir Qureshi/AFP/Getty Images
26–27 Used with kind permission of Sverrir Þórolfsson
32–33 Brynjar Gauti/AP/Press Association Images
34 Bernhard Edmaier/Science Photo Library
35 NASA/Earth Observatory
36 MODIS/NASA
37 EO-1/NASA
38 MODIS/NASA
39 Reuters/Handout
40–41 Sadatsugu Tomizawa/AFP/Getty Images
42–43 NOAA/Science Photo Library
44–45 Images courtesy of GeoEye
46 Toru Yamanaka, KimJae-Hwan/AFP/Getty Imges
47 Toru Yamanaka, Philippe Lopez/AFP/Getty Imges
48 NOAA/Science Photo Library
49 NASA Earth Observatory image by Robert Simmon, using data from the NASA/GSFC/METI/ERSDAC/JAROS, and U.S./Japan ASTER Science Team
50–51 IKONOS images © CRISP 2004
52–53 Jefferey Liao/Shutterstock
55 NOAA/Science Photo Library
56 **top** PA/PA/EMPICS
56 **bottom** John Giles/PA/EMPICS
57 John Giles/PA/EMPICS
58 Images courtesy of GeoEye
59 NASA images created by Jesse Allen, using data provided courtesy of GSFC/METI/ERSDAC/JAROS, and the U.S./Japan ASTER Science Team.
60 NASA/Earth Observatory
61 © Belcastro Antonietta/epa/Corbis
62–63 Hedgehog House/Nick Groves
65 MODIS/NASA
68–69 Shutterstock/Robert Ranson
73 IKONOS images courtesy of GeoEye
74 NASA GFSC GOES Project
75 Image courtesy of the NOAA Hurricane Irene Project
76–77 MODIS/NASA
78 Image Science and Analysis Laboratory, NASA-Johnson Space Center. The Gateway to Astronaut Photography of Earth.
78 **inset top** NASA Earth Observatory
79 Bradley Kanaris/Getty Images
80–81 Shutterstock/Photography Perspectives - Jeff Smith
84–85 Michel Gunther/Robert Harding
86 **top** NOAA-NASA GOES Project
86 **bottom** NASA/Earth Observatory
87 AFP/Getty Images
88–89 Benjamin Krain/Getty Images
90–91 Shutterstock/André Klaassen
93 MODIS/NASA
94 Jan Kratochvila/Shutterstock
95 Jack Reynolds/Getty
96–97 NOAA/Department of Commerce
98–99 Daniel J Bryant/Getty
100–101 Shutterstock/Jason Maehl
103 Reuters/Denis Bailhouse
104–105 NASA/GSFC
106–107 NASA/GSFC
108 MODIS/NASA
109 **top** Carl de Souza/AFP/Getty Images
109 **bottom** © Jonathan Brady/epa/Corbis
110 European Space Agency, ESA
111 **top** imagestalk/Shutterstock.com
111 **bottom** alessandro0770/Shutterstock.com
113 MODIS/NASA
116–117 Image Science and Analysis Laboratory, NASA-Johnson Space Center. The Gateway to Astronaut Photography of Earth.
120 IKONOS images courtesy of GeoEye
121 NASA images by Robert Simmon, based on data from the NASA/GSFC/METI/ERSDAC/JAROS, and U.S./Japan ASTER Science Team
122 **top** Robert Harding Picture Library Ltd/Alamy
122 **bottom** © Steve Vidler/Photoshot
123 NASA/Earth Observatory
124–125 NASA Earth Observatory image created by Robert Simmon and Jesse Allen, using Landsat data provided by the United States Geological Survey.
128–129 MODIS/NASA
130 NASA/GSFC/METI/Japan Space Systems, and U.S./Japan ASTER Science Team
131 Image Science and Analysis Laboratory, NASA-Johnson Space Center. The Gateway to Astronaut Photography of Earth.
132–133 Image courtesy of GeoEye
139 AFP/Getty Images
140 © Bettman/CORBIS
141 Bloomberg via Getty Images
142–143 NASA/Earth Observatory
144–145 NASA/GSFC/METI/Japan Space Systems, and U.S./Japan ASTER Science Team
146 Mark Edwards/Robert Harding
147 Alexandre Meneghini/Empics
149 Image courtesy Julia Bucknall/World Bank
152–153 Dhoxax/Shutterstock
158–159 NASA/Earth Observatory
160 **top** Greenpeace/Daniel Beltrá
160 **bottom** Riccardo Pravettoni http://www.grida.no/photolib/detail/geometrical-scars-on-the-land-brazil_b24c
161 Riccardo Pravettoni http://www.grida.no/photolib/detail/aerial-view-of-the-amazon-jungle_c9e8
162–163 S.Rocha/UNEP/Robert Harding
164 **top** Peter Prokosch http://www.grida.no/photolib/detail/deforested-landscape-in-central-south-madagascar_1bc5
164 **bottom** Image Science and Analysis Laboratory, NASA-Johnson Space Center. The Gateway to Astronaut Photography of Earth.
165 **top** Peter Prokosch http://www.grida.no/photolib/detail/burning-rainforest-on-sumatra-to-make-space-for-palm-oil-plantations-indonesia_8277
165 **bottom** Romeo Gacad/AFP/Getty Images
166–167 Woodfall/Photoshot
170 **top** Mike Hewitt/Getty Images
170 **bottom** Graeme Brown/Maritime New Zealand via Getty Images
171 AFP/Getty Images
172 **top** AlamyCelebrity/Alamy
172 **bottom** MODIS/NASA
173 **top** Greenpeace/Daniel Beltrá
173 **bottom** © Jose Luís Magana/Greenpeace
174–175 NASA/Earth Observatory
176 NASA/Earth Observatory
177 © Peter Somogyi-Tóth/Greenpeace
178 **top** Georgette Douwma/Science Photo Library
178 **bottom** Martha Holmes/Naturepl.com
179 **top** Michael Pitts/Naturepl.com
179 **bottom** Greenpeace/Marco Care
180 MODIS/NASA
181 **top** IKONOS image courtesy of GeoEye
181 **bottom** NASA/Earth Observatory
184–185 Armin Rose/Shutterstock
186 NASA Earth Observatory/Robert Simmon. Data Sources: NASA Goddard Institute for Space Studies, NOAA National Climatic Data Center, Met Office Hadley Centre/Climatic Research Unit, and the Japanese Meteorological Agency.
188 NASA/Goddard Space Flight Center Scientific Visualization Studio
189 NSIDC Boulder, Colorado, USA
190 **top** NASA/Earth Observatory
190 Images of Antarctic ice sheets. Boulder, Colorado, USA: National Snow and Ice Data Center. Digital media. http://nsidc.org/data/iceshelves_images/ **middle** 10th April 2009, **bottom** 7th January 2010
191 NASA/Earth Observatory
192–193 MODIS/NASA
194 Image by the NASA Scientific Visualization Studio based on data from the Special Sensor Microwave Imager/Sounder (SSMIS) of the Defense Meteorological Satellite Program (DMSP).
195 Image courtesy Andy Mahoney and the National Snow and Ice Data Center, University of Colorado, Boulder, CO, USA
196 **top** © B&C Alexander/ArcticPhoto
196 **bottom** Lawrence Hislop http://www.grida.no/photolib/detail/aerial-view-of-shishmaref_3aaa
197 **top** © B&C Alexander/ArcticPhoto
197 **bottom** Lawrence Hislop http://www.grida.no/photolib/detail/aerial-view-of-the-old-runway-now-used-for-housing_3926
198 NASA/Earth Observatory
199 **top** © Alaska Stock/Corbis
199 **bottom** Jean-Paul Ferrero/Auscape/Minden/Corbis
200–201 Shutterstock/javarman
204 Jürg Alean, Eglisau, Switzerland http://www.swisseduc.ch
205 Print 1427. Detroit Publishing Company, 1905.
206 **top** E.C. Stebinger, courtesy of Glacier National Park Archives
206 **bottom** Lisa McKeon/USGS
207 **left** T.J. Hileman, courtesy of Glacier National Park Archives
207 **right** Lindsey Bengtson/USGS
208 **top** Shutterstock/Peter Wolinga
208 **bottom** Shutterstock/tomtsya
209 NASA/Earth Observatory
210–211 NASA/Earth Observatory
212 **images 1–8** NSIDC, Boulder, Colorado, USA
212 **image 9** en:user:dramatic/Creative Commons Attribution-Share Alike 3.0 Unported license
213 Bart Pro/Alamy
214–215 Shutterstock/Antonio S.
216 **top** Colorado Center for Astrodynamics Research at the University of Colorado at Boulder
216 **bottom** IPCC 2007. WGI AR4
218 Sutterstock/Eugene Moglinikov
219 Alvaro German Vilela/Shutterstock.com
220 NASA/GSFC/METI/Japan Space Systems, and U.S./Japan ASTER Science Team
221 **top** ESA Envistat
221 **bottom** Chris 73/Wikimedia Commons, under the creative commons cc-by-sa 3.0 license.
222 IKONOS image courtesy of GeoEye
223 **top** Shahee Ilyas
223 **bottom** shutterstock/poakun
224 Matty Symons/Shutterstock
225 NASA Earth Observatory
228–229 Shutterstock/Pichugin Dmitry
232–233 Voltchev/UNEP/Robert Harding
234–235 George Steinmetz/Science Photo Library
236 Image courtesy of MODIS Rapid Response Team at NASA/GSFC
237 George Steinmetz/Science Photo Library
238 Jorgen Schytte/Robert Harding
239 © Redlink/Corbis
240–241 Jeff Strickler/Shutterstock
244–245 NASA/Earth Observatory
246–247 NASA/Earth Observatory
248 AFP/Getty Images
249 **top** 4780322454/Shutterstock
249 **bottom** Viacheslav Lopatin/Shutterstock.com
250–251 A. Vogler/Shutterstock
254–255 Images reproduced by kind permission of UNEP
255 **inset** NASA/Earth Observatory
256 **1973, 1986, 2001** USGS, EROS Data Center, Sioux Falls, SD
256 **2005** MODIS/NASA
257 **top** NASA/Earth Observatory
257 **bottom** Vyacheslav Oseledko/AFP/Getty Images
258–259 NASA/Earth Observatory
260–261 NASA/Earth Observatory
263 NASA Earth Obervatory
266–267 Shutterstock/Angela Crowley
270–271 Images reproduced by kind permission of UNEP
271 **inset** NASA Earth Observatory
272–273 © Parks Victoria/Handout/Reuters/CORBIS
274 © Gary Braasch Environmental Photography
275 **top** Woodfall/Photoshot
275 **bottom** © Andrew Stacey
276–277 Scyscan/Corbis
282–283 NASA Earth Observatory
284–285 NASA Earth Observatory
286–287 MODIS/NASA
288–289 Reuters/Handout
291 **top** © Paul Jenkin/Rich Reid Photography
291 **bottom** © Rich Reid Photography